This is an important collection of papers on the foundations of probability that will be of value to philosophers of science, mathematicians, statisticians, psychologists, and educators.

The collection falls into three parts. Part I comprises five papers on the axiomatic foundations of probability. Part II contains seven articles on probabilistic causality and quantum mechanics, with an emphasis on the existence of hidden variables. The third part consists of a single extended essay applying probabilistic theories of learning to practical questions of education: it incorporates extensive data analysis.

Patrick Suppes in one of the world's foremost philosophers in the area of probability and has made many contributions to both the theoretical and the practical side of education. The statistician Mario Zanotti is a long-time collaborator.

Foundations of probability with applications

Foundations of probability with applications

Selected papers, 1974–1995

Patrick Suppes

Stanford University

Mario Zanotti

Computer Curriculum Corporation

CAMBRIDGE
UNIVERSITY PRESS

Published by the Press Syndicate of the University of Cambridge
The Pitt Building, Trumpington Street, Cambridge CB2 1RP
40 West 20th Street, New York, NY 10011-4211, USA
10 Stamford Road, Oakleigh, Melbourne 3166, Australia

First published 1996

Library of Congress Cataloging-in-Publication Data

Suppes, Patrick, 1922–

Foundations of probability with applications: selected papers
1974–1995 / Patrick Suppes, Mario Zanotti.

p.   cm. – (Cambridge Studies in probability, induction, and decision theory)

ISBN 0-521-43012-7. – ISBN 0-521-56835-8 (pbk.)

1. Probabilities.   I. Zanotti, Mario.   II. Title.   III. Series.
QA273.18.S86   1996
519.2 – dc20   96-7895
                                                          CIP

A catalog record for this book is available from the British Library.

ISBN 0-521-43012-7 Hardback
ISBN 0-521-56835-8 Paperback

Transferred to digital printing 2003

Contents

Part III: Applications in education

Preface

The joint articles that we have written over the past twenty years fall into three different areas and consequently there is a natural division of this collection of our papers into three parts. Part I is concerned with our work in the foundations of probability, Part II with causality and quantum mechanics, and Part III with application of probabilistic models in education.

The five papers in Part I represent our joint efforts to clarify and extend the qualitative foundations of probability first given impetus in the important work of Bruno de Finetti. Both of us, but separately over a good many years, had extensive conversations with de Finetti about the foundations of probability, and several of our papers were inspired by questions he raised. In fact, the first two articles on necessary and sufficient conditions for existence of a strictly agreeing measure of a qualitative probability ordering were a response to de Finetti's early qualitative axioms and the subsequent search for necessary and sufficient axioms. The important point about our work is that in order to get simple axioms we had to go beyond the usual event structure of a probability space to elementary random variables. But we felt at the time, and continue to feel, that this is an extension that is very much in de Finetti's own line of thought, as reflected for example, in his introduction at an early stage of random quantities without an underlying probability space to represent them. It is also worth noting that characterizing the expectation of elementary random variables axiomatically builds on a long tradition in the theory of probability to define probability in terms of expectation. This is already to be found in Bayes's eighteenth century treatise.

Another direction of representation is to generalize from probabilities that subjectively seem sometimes difficult to manage to the weaker concept of having upper and lower probabilities. The third and fourth papers deal with such problems, and particularly the fourth paper is related to questions raised by de Finetti. In particular we examine in this article the conditions on upper and lower probabilities for them to imply the existence of probabilities. The fifth and final paper in Part I addresses a somewhat different but related problem, namely, that of giving random-variable rather than numerical representation for extensive quantities in the theory of measurement. Here the methods are somewhat different and in fact we use in a central way the well-known Hausdorf moment theorem to give qualitative conditions for numerical random variables whose distributions have finite support, that is, are defined on a bounded set. The central problem addressed in this paper is very much related to the long history of the study of the theory of error in probability theory going back to the early work of Simpson in the eighteenth century and later work by many others. This problem of finding appropriate qualitative axioms for errors or variability in measurements remains one of the least satisfactorily solved problems in the foundations of measurement.

Part II contains seven papers on the foundations of the theory of probabilistic causality and more particularly on the foundations of quantum mechanics. The first three papers deal directly with problems in quantum mechanics, the first with the stochastic incompleteness of quantum mechanics looked at from the standpoint of stochastic processes, the second and third with the problems that arise in connection with hidden-variable theories and in particular with the problem of such theories in the context of Bell's theorem. Our 1976 paper on these matters, which is the second paper in this section, was among the earliest to introduce in a formal way precise probabilistic statements about the independence conditions required for the derivation of Bell's theorem.

In the third paper we move away from specific questions in quantum mechanics to showing the impossibility of hidden variables when we impose natural conditions such as exchangeability and symmetry. It is important for positive scientific work regarding hidden variables to understand that rather natural conditions are sufficient to show that hidden variables cannot exist. The fourth paper moves in the opposite direction. Here we show that without such conditions probabilistic explanations in terms of hidden variables, in fact deterministic ones, are always possible whenever the observable random variables have a joint probability distribution. This is an interesting twist on the old and mistaken tale of the conflict between determinism and probability. Here they go hand in hand. Deterministic hidden variables can be found if and only if the phenomenological observables have a joint probability distribution. In the search for hidden variables that have conceptual meaning we need to impose much stronger conditions of the sort to be found in the third paper in this section. As the fourth paper shows, without such conditions a mathematical construction of hidden variables can always be given, even if this mathematical construction does not have any direct conceptual interpretation in terms of the scientific framework involved.

The fifth paper reviews and extends our earlier work on probabilistic causality and symmetry with special reference to quantum mechanics. The sixth paper returns to hidden variables in the context of Bell's theorem and gives a necessary condition for existence of a hidden variable when the number of hidden variables is greater than four. Recall that in Bell's theorem only four observables make up the inequalities, but it is natural to go beyond four and to ask more general questions. Finally, in the seventh paper in this section, written to celebrate the 90th birthday of Karl Popper, we show that if we weaken the conditions in the Bell-type situation to the existence, not of a probability distribution, but of an upper probability measure, then such a measure can always be found that is also compatible with the results of quantum mechanics. However, as we show in the paper, this upper probability measure is not monotonic, which means that there are events A and B such that even though A must occur if B occurs, the upper probability measure assigns a lower measure to A than to B. We extend in that framework our earlier results on causality to the existence of common causes in the framework of such upper measures. The results turn out to be rather satisfactory from the standpoint of their neatness of formulation, but whether or not they will end up having any application scientifically is yet to be seen.

Part III on applications in education of probabilistic models contains only one paper. It has not been previously published, but in fact over the past twenty years we have probably devoted as much of our joint effort to this work as to any other. Much of what we have done remains unpublished, although our first publication in this domain appeared in 1976 and, in fact, even earlier as a technical report in 1973.

References to our earlier work in education are given at the end of the long article that constitutes Part III. This paper deals with work that we are continuing and will continue beyond the point at which the present volume appears, namely, work on stochastic models of mastery learning. Here we try to give a sample of the kind of detailed empirical work we think is possible in education in the application of specific probabilistic models of learning to actual instruction when it takes place in the framework of computer-based education. It would be our hope that there will be in the future much further development of the kind of models we describe and develop, which are, as can be seen in the complete formulation at the end of the paper, rather intricate.

Acknowledgments for permission to reproduce the various articles are given at the bottom of the first page of each paper, but thanks are extended here to the many editors and publishers who generously agreed to publication. Finally, we acknowledge the help and patience of Laura von Kampen in preparing this volume for publication.

Patrick Suppes
Mario Zanotti

Stanford, California

I

Foundations of probability

1

Necessary and sufficient conditions for existence of a unique measure strictly agreeing with a qualitative probability ordering

1. CONCEPTUAL BACKGROUND

Let Ω be a nonempty set and let $\mathcal{F}$ be an algebra of events on Ω, i.e., an algebra of sets on Ω. Let $\succcurlyeq$ be a qualitative ordering on $\mathcal{F}$. The interpretation of $A \succcurlyeq B$ for two events A and B is that A is *at least as probable* as B. A (finitely additive) probability measure P on $\mathcal{F}$ is *strictly agreeing* with the relation $\succcurlyeq$ if and only if, for any two events A and B in $\mathcal{F}$,

$$P(A) \geqslant P(B) \text{ iff } A \succcurlyeq B.$$

A variety of conditions that guarantee the existence of a strictly agreeing measure is known. Without attempting a precise classification, the sets of conditions are of the following sorts: (i) sufficient but not necessary conditions for existence of a unique measure when the algebra of events is infinite (Koopman, 1940; Savage, 1954; Suppes, 1956); (ii) sufficient but not necessary conditions for uniqueness when the algebra of events is finite or infinite (Luce, 1967); sufficient but not necessary conditions for uniqueness when the algebra of events is finite (Suppes, 1969); (iv) necessary and sufficient conditions for existence of a not necessarily unique measure when the algebra of events is finite (Kraft, Pratt, & Seidenberg, 1959; Scott, 1964; Tversky, 1967). A rather detailed discussion of these various sets of conditions is to be found in Chapters 5 and 9 of Krantz, Luce, Suppes, and Tversky (1971).

The difficulties of giving reasonably simple conditions in terms of the qualitative ordering of events are exemplified by Luce's axiom, which is weaker than Koopman's equipartition axiom or Savage's related but somewhat stronger axiom. Luce's axiom is the following (Krantz *et al.*, 1971, p. 207):

Reprinted from *Journal of Philosophical Logic* 5 (1976), 431–438.

For any events A, B, C, and D such that $A \cap B = \emptyset$, $A \succ C$, and $B \succcurlyeq D$, there exist events C', D', and E such that

(i) $E \approx A \cup B$;
(ii) $C' \cap D' = \emptyset$;
(iii) $C' \cup D' \subset E$;
(iv) $C' \approx C$ and $D' \approx D$.

Here $\succ$ is the strict ordering relation and $\approx$ the equivalence relation defined in terms of the weak ordering $\succcurlyeq$. The meaning of this axiom is complex and not easy to state in words. As we search for weaker axioms, closer to being necessary and not merely sufficient, the situation seems likely to get worse. The moral of the effort is that events are the wrong objects to consider. Some slightly richer concept is needed. Extension from one set of objects to a larger and richer set is a characteristic move in mathematics. The most familiar examples are extension of the rational numbers to the real numbers and extension of the real numbers to the complex numbers. As Georg Kreisel has emphasized in several conversations, the introduction of auxiliary concepts is an indispensable practical move in solving significant problems in many domains of mathematics and science.

The main result of this article exemplifies how easily simplification can follow from the introduction of auxiliary concepts. In the present case the move is from an algebra of events to an algebra of extended indicator functions for the events. By this latter concept we mean the following. As before, let Ω be the set of possible outcomes and let $\mathcal{F}$ be an algebra of events on Ω, i.e., $\mathcal{F}$ is a nonempty family of subsets of Ω, and is closed under complementation and union, i.e., if A is in $\mathcal{F}$, $\neg A$, the complement of A with respect to Ω, is in $\mathcal{F}$, and if A and B are in $\mathcal{F}$ then $A \cup B$ is in $\mathcal{F}$. Let A^c be the indicator function (or characteristic function) of event A. This means that A^c is a function defined on Ω such that for any ω in Ω,

$$A^c(\omega) = \begin{cases} 1 \text{ if } \omega \in A \\ 0 \text{ if } \omega \notin A. \end{cases}$$

The algebra $\mathcal{F}^*$ of *extended* indicator functions relative to $\mathcal{F}$ is then just the smallest semigroup (under function addition) containing the indicator functions of all events in $\mathcal{F}$. In other words, $\mathcal{F}^*$ is the intersection of all sets with the property that if A is in $\mathcal{F}$ then A^c is in $\mathcal{F}^*$, and if A^* and B^* are in $\mathcal{F}^*$, then $A^* + B^*$ is in $\mathcal{F}^*$. It is easy to show that any function A^* in $\mathcal{F}^*$ is an integer-valued function defined on Ω. It is the extension from indicator functions to integer-valued functions that justifies calling the elements of $\mathcal{F}^*$ extended indicator functions.

The qualitative probability ordering must be extended from $\mathcal{F}$ to $\mathcal{F}^*$, and the intuitive justification of this extension must be considered. Let A^* and B^* be

4

two extended indicator functions in $\mathcal{F}^*$. Then, to have $A^* \succeq B^*$ is to have the expected value of A^* equal to or greater than the expected value of B^*. As should be clear, extended indicator functions are just random variables of a restricted sort. The qualitative comparision is now not one about the probable occurrences of events, but about the expected value of certain restricted random variables. The indicator functions themselves form, of course, a still more restricted class of random variables, but qualitative comparison of their expected values is conceptually identical to qualitative comparison of the probable occurrences of events.

There is more than one way to think about the qualitative comparison of the expected value of extended indicator functions, and so it is useful to consider several examples.

(i) Suppose Smith is considering two locations to fly to for a weekend vacation. Let A_i be the event of sunny weather at location i and B_i be the event of warm weather at location i. The qualitative comparision Smith is interested in is the expected value of $A_1^c + B_1^c$ *versus* the expected value of $A_2^c + B_2^c$. It is natural to insist that the utility of the outcomes has been too simplified by the sums $A_i^c + B_i^c$. The proper response is that the expected values of the two functions are being compared as a matter of belief, not value or utility. Thus it would seem quite natural to bet that the expected value of $A_1^c + B_1^c$ will be greater than that of $A_2^c + B_2^c$, no matter how one feels about the relative desirability of sunny *versus* warm weather. Put another way, within the context of decision theory, extended indicator functions are being used to construct the subjective probability measure, not the measurement of utility. In this context it is worth recalling the importance of certain special decision functions – the gambles – in Savage's theory.

(ii) Consider a particular population of n individuals, numbered $1, \ldots, n$. Let A_i be the event of individual i going to Hawaii for a vacation this year, and let B_i be the event of individual i going to Acapulco. Then define

$$A^* = \sum_{i=1}^{n} A_i^c \qquad \text{and} \qquad B^* = \sum_{i=1}^{n} B_i^c.$$

Obviously A^* and B^* are extended indicator functions – we have left implicit the underlying set Ω. It is meaningful and quite natural to qualitatively compare the expected values of A^* and B^*. Presumably such comparisons are in fact of definite significance to travel agents, airlines, and the like.

We believe that such qualitative comparisons of expected value are natural in many other contexts as well. What the main theorem of this article shows is that very simple necessary and sufficient conditions on the qualitative comparison of extended indicator functions guarantee existence of a strictly agreeing, finitely additive measure, whether the set Ω of possible outcomes is finite or

5

infinite. Moreover, when it is required that the measure also be an expectation function for the extended indicator functions, it is unique. The proof of the theorem, it should be mentioned, depends directly upon the theory of extensive measurement developed in Chapter 3 of Krantz *et al.* (1971).

2. FORMAL DEVELOPMENTS

The axioms are embodied in the definition of a qualitative algebra of extended indicator functions. Several points of notation need to be noted. First, Ω^c and $\emptyset^c$ are the indicator or characteristic functions of the set Ω of possible outcomes and the empty set $\emptyset$, respectively. Second, the notation nA^* for a function in $\mathcal{F}^*$ is just the standard notation for the (functional) sum of A^* with itself n times. Third, the same notation is used for the ordering relation on $\mathcal{F}$ and $\mathcal{F}^*$, because the one on $\mathcal{F}^*$ is an extension of the one on $\mathcal{F}$: for A and B in $\mathcal{F}$,

$$A \succcurlyeq B \text{ iff } A^c \succcurlyeq B^c.$$

Finally, the strict ordering relation $\succ$ is defined in the usual way: $A^* \succ B^*$ iff $A^* \succcurlyeq B^*$ and not $B^* \succcurlyeq A^*$.

DEFINITION *Let Ω be a nonempty set, let $\mathcal{F}$ be an algebra of sets on Ω, and let $\succcurlyeq$ be a binary relation on $\mathcal{F}^*$, the algebra of extended indicator functions relative to $\mathcal{F}$. Then the qualitative algebra $(\Omega, \mathcal{F}^*, \succcurlyeq)$ is qualitatively satisfactory if and only if the following axioms are satisfied for every A^*, B^*, and C^* in $\mathcal{F}^*$:*

Axiom 1. The relation $\succcurlyeq$ is a weak ordering of $\mathcal{F}^*$;
Axiom 2. $\Omega^c \succcurlyeq \emptyset^c$;
Axiom 3. $A^* \succcurlyeq \emptyset^c$;
Axiom 4. $A^* \succcurlyeq B^*$ iff $A^* + C^* \succcurlyeq B^* + C^*$;
Axiom 5. If $A^* \succ B^*$ then for every C^* and D^* in $\mathcal{F}^*$ there is a positive integer n such that

$$nA^* + C^* \succcurlyeq nB^* + D^*.$$

These axioms should seem familiar from the literature on qualitative probability. Note that Axiom 4 is the additivity axiom that closely resembles de Finetti's additivity axiom for events: *If $A \cap C = B \cap C = \emptyset$, then $A \succcurlyeq B$ iff $A \cup C \succcurlyeq B \cup C$.* As we move from events to extended indicator functions, functional addition replaces union of sets. What is formally of importance about this move is seen already in the exact formulation of Axiom 4. The additivity of the extended indicator functions is unconditional – there is no restriction corresponding to $A \cap C = B \cap C = \emptyset$. The absence of this restriction has far-reaching formal consequences in permitting us to apply without any real

6

modification the general theory of extensive measurement. Axiom 5 has, in fact, the exact form of the Archimedean axiom used in Krantz *et al.* (1971, p. 73) in giving necessary and sufficient conditions for extensive measurement. Discussion of why the formally simpler axiom – if $A^* \succeq B^* \succ \emptyset^c$ then there is an n such that $nB^* \succeq A^*$ – is not precisely satisfactory in giving necessary and sufficient axioms will be found there.

We are now in a position to formulate the theorem that is paraphrased in the title of the article.

THEOREM *Let Ω be a nonempty set, let $\mathcal{F}$ be an algebra of sets on Ω, and let $\succeq$ be a binary relation on $\mathcal{F}$. Then a necessary and sufficient condition that there exist a strictly agreeing probability measure on $\mathcal{F}$ is that there is an extension of $\succeq$ from $\mathcal{F}$ to $\mathcal{F}^*$ such that the qualitative algebra of extended indicator functions $(\Omega, \mathcal{F}^*, \succeq)$ is qualitatively satisfactory. Moreover, if $(\Omega, \mathcal{F}^*, \succeq)$ is qualitatively satisfactory, then there is a unique strictly agreeing expectation function on $\mathcal{F}^*$ and this expectation function generates a unique strictly agreeing probability measure on $\mathcal{F}$.*

Proof. As already indicated, the main tool used in the proof is from the theory of extensive measurement: necessary and sufficient conditions for existence of a numerical representation, as given in Krantz *et al.* (1971, pp. 73–74, Theorem 1). More precisely, let A be a nonempty set, $\succeq$ a binary relation on A, and $\circ$ a binary operation closed on A. Then there exists a numerical function φ on A unique up to a positive similarity transformation (i.e., multiplication by a positive real number) such that for a and b in A

(i) $a \succeq b$ iff $\varphi(a) \geqslant \varphi(b)$,
(ii) $\varphi(a \circ b) = \varphi(a) + \varphi(b)$

if and only if the following four axioms are satisfied for all a, b, c, and d in A:

E1: The relation $\succeq$ is a weak ordering of A;
E2: $a \circ (b \circ c) \approx (a \circ b) \circ c$, where $\approx$ is the equivalence relation defined in terms of $\succeq$;
E3: $a \succeq b$ iff $a \circ c \succeq b \circ c$ iff $c \circ a \succeq c \circ b$;
E4: If $a \succ b$ then for any c and d in A there is a positive integer n such that $na \circ c \succeq nb \circ d$, where na is defined inductively.

It is easy to check that qualitatively satisfactory algebras of extended indicator functions as defined above satisfy these four axioms for extensive measurement structures. First, we note that functional addition is closed on $\mathcal{F}^*$. Second, Axiom 1 is identical to E1. Extensive Axiom E2 follows immediately from the associative property of numerical functional addition: For any A^*, B^*, and C^*

7

in $\mathcal{F}^*$

$$A^* + (B^* + C^*) = (A^* + B^*) + C^*,$$

and so we have not just equivalence but identity. Axiom E3 follows from Axiom 4 and the fact that numerical functional addition is commutative. Finally, E4 follows from the essentially identical Axiom 5.

Thus, for any qualitatively satisfactory algebra $(\Omega, \mathcal{F}^*, \succcurlyeq)$ we can infer there is a numerical function φ on Ω such that for A^* and B^* in $\mathcal{F}^*$

(i) $A^* \succcurlyeq B^*$ iff $\varphi(A^*) \succcurlyeq \varphi(B^*)$

(ii) $\varphi(A^* + B^*) = \varphi(A^*) + \varphi(B^*)$,

and φ is unique up to a positive similarity transformation.

Second, since for every A^* in $\mathcal{F}^*$

$$A^* + \emptyset^c = A^*,$$

we have at once from (ii)

$$\varphi(\emptyset^c) = 0.$$

Since $\Omega^c \succ \emptyset^c$ by Axiom 2, we can choose

$$\varphi(\Omega^c) = 1,$$

and thus have a standard (unique) expectation function E for extended indicator functions:

(i) $E(\emptyset^c) = 0$

(ii) $E(\Omega^c) = 1$

(iii) $E(A^* + B^*) = E(A^*) + E(B^*)$.

But such an expectation function for $\mathcal{F}^*$ defines a unique probability measure P on $\mathcal{F}$ when it is restricted to the indicator functions in $\mathcal{F}^*$, i.e., for A in $\mathcal{F}$, we define

$$P(A) = E(A^c).$$

Thus the axioms are sufficient, but it is also obvious that the only axioms, Axioms 2 and 3, that go beyond those for extensive structures are also necessary for a probabilistic representation.

From the character of extended indicator functions, it is also clear that for each probability measure there is a unique extension of the qualitative ordering from $\mathcal{F}$ to $\mathcal{F}^*$.

The proof just given, even more than the statement of the theorem itself, shows what subset of random variables defined on a probability space suffices to determine the probability measure in a natural way. Our procedure has been to axiomatize in qualitative fashion the expectation of the extended indicator functions. There was no need to consider all random variables, and, on the other hand, the more restricted set of indicator functions raises the same axiomatic difficulties confronting the algebra of events. The natural sequence of developments is then:

(i) Axiomatize qualitatively the expectation of extended indicator functions;
(ii) Use this expectation to determine the unique probability measure;
(iii) Use the probability measure to determine the expectation of all random variables defined on the space.

The great simplicity of our necessary and sufficient conditions supports the intuitive correctness of this sequence of developments, although, as far as we know, the theory of extended indicator functions has not previously played a noticeable role in the theory of qualitative probability.

NOTE

The ideas and results contained in this article were first presented at the Stanford Seminar on Foundations of Probability, Winter Quarter, 1976. We benefited considerably from comments and criticisms of several members of the seminar, especially Thomas Cover.

REFERENCES

Koopman, B. O.: 1940, 'The Bases of Probability', *Bulletin of the American Mathematical Society*, **46**, 763–774.
Kraft, C. H., Pratt, J. W. and Seidenberg, A.: 1959, 'Intuitive Probability on Finite Sets', *Annals of Mathematical Statistics*, **30**, 408–419.
Krantz, D. H., Luce, R. D., Suppes, P. and Tversky, A.: 1971, *Foundations of Measurement* (Vol. 1), Academic Press, New York.
Luce, R. D.: 1967, 'Sufficient Conditions for the Existence of a Finitely Additive Probability Measure', *Annals of Mathematical Statistics*, **38**, 780–786.
Savage, L. J.: 1954, *The Foundations of Statistics*, Wiley, New York.
Scott, D.: 1964, 'Measurement Models and Linear Inequalities', *Journal of Mathematical Psychology*, **1**, 233–247.
Suppes, P.: 1956, 'The Role of Subjective Probability and Utility in Decision-making', in J. Neyman (Ed.), *Proceedings of the Third Berkeley Symposium of Mathematical Statistics and Probability* (Vol. 5), University of California Press, Berkeley.
Suppes, P.: 1969, *Studies in the Methodology and Foundations of Science*, Reidel, Dordrecht.
Tversky, A.: 1967, 'Additivity, Utility, and Subjective Probability', *Journal of Mathematical Psychology*, **4**, 175–201.

2

Necessary and sufficient qualitative axioms for conditional probability

1. INTRODUCTION

In a previous paper (Suppes and Zanotti, 1976) we gave simple necessary and sufficient qualitative axioms for the existence of a unique expectation function for the set of extended indicator functions. As we defined this set of functions earlier, it is the closure of the set of indicator functions of events under function addition. In the present paper we extend the same approach to conditional probability. One of the more troublesome aspects of the qualitative theory of conditional probability is that $A \mid B$ is not an object – in particular it is not a new event composed somehow from events A and B. Thus the qualitative theory rests on a quaternary relation $A \mid B \succcurlyeq C \mid D$, which is read: event A given event B is at least as probable as event C given event D. There have been a number of attempts to axiomatize this quaternary relation (Koopman, 1940a, 1940b; Aczél, 1961, 1966, p. 319; Luce, 1968; Domotor, 1969; Krantz *et al.*, 1971; and Suppes, 1973). The only one of these axiomatizations to address the problem of giving necessary and sufficient conditions is the work of Domotor, which approaches the subject in the finite case in a style similar to that of Scott (1964).

By using indicator functions or, more generally, extended indicator functions, the difficulty of $A \mid B$ not being an object is eliminated, for $A^c \mid B$ is just the indicator function of the set A restricted to the set B, that is, $A^c \mid B$ is a partial function whose domain is B. In similar fashion if X is an extended indicator function, $X \mid A$ is that function restricted to the set A. The use of such partial functions requires care in formulating the algebra of functions in which we are interested, for functional addition $X \mid A + Y \mid B$ will not be well defined when

Reprinted from Z. Wahrscheinlichkeitstheorie verw. Gebiete **60** (1982), 163–169.

$A \neq B$ but $A \cap B \neq \emptyset$. Thus, to be completely explicit we begin with a nonempty set Ω, the probability space, and an algebra $\mathcal{F}$ of events, that is, subsets of Ω, with it understood that $\mathcal{F}$ is closed under union and complementation. Next we extend this algebra to the algebra $\mathcal{F}^*$ of extended indicator functions, that is, the smallest semigroup (under function addition) containing the indicator functions of all events in $\mathcal{F}$. This latter algebra is now extended to include as well all partial functions on Ω that are extended indicator functions restricted to an event in $\mathcal{F}$. We call this algebra of partial extended indicator functions $\mathcal{RF}^*$, or, if complete explicitness is needed, $\mathcal{RF}^*(\Omega)$. From this definition it is clear that if $X \mid A$ and $Y \mid B$ are in $\mathcal{RF}^*$, then

(i) If $A = B, X \mid A + Y \mid B$ is in $\mathcal{RF}^*$.

(ii) If $A \cap B = \emptyset, X \mid A \cup Y \mid B$ is in $\mathcal{RF}^*$.

In the more general setting of decision theory or expected utility theory there has been considerable discussion of the intuitive ability of a person to directly compare his preferences or expectations of two decision functions with different domains of restriction. Without reviewing this literature, we do want to state that we find no intuitive general difficulty in making such comparisons. Individual cases may present problems, but not necessarily because of different domains of definition. In fact, we believe comparisons of expectations under different conditions is a familiar aspect of ordinary experience. In the present setting the qualitative comparison of restricted expectations may be thought of as dealing only with beliefs and not utilities. The fundamental ordering relation is a weak ordering $\succeq$ of $\mathcal{RF}^*$ with strict order $\succ$ and equivalence $\sim$ defined in the standard way.

The axioms we give are strong enough to prove that the probability measure constructed is unique when it is required to cover expectation of random variables. It is worth saying something more about this problem of uniqueness. The earlier papers mentioned have all concentrated on the existence of a probability distribution, but from the standpoint of a satisfactory theory it seems obvious for many different reasons that one wants a unique distribution. For example, if we go beyond properties of order and have uniqueness only up to a convex polyhedron of distributions, as is the case with Scott's axioms for finite probability spaces, we are not able to deal with a composite hypothesis in a natural way, because the addition of the probabilities is not meaningful.

2. Axioms

We incorporate our axioms in the usual form of a definition.

DEFINITION *Let Ω be a nonempty set, let $\mathcal{RF}^*(\Omega)$ be an algebra of partial extended indicator functions, and let $\succeq$ be a binary relation on $\mathcal{RF}^*$. Then the*

11

structure $(\Omega, \mathcal{RF}^*, \succcurlyeq)$ *is a* partial qualitative expectation structure *if and only if the following axioms are satisfied for every X and Y in $\mathcal{F}^*$ and every A, B and C in $\mathcal{F}$ with $A, B \succ \emptyset$:*

Axiom 1. The relation $\succcurlyeq$ is a weak ordering of $\mathcal{RF}^*$;

Axiom 2. $\Omega^c \succ \emptyset^c$;

Axiom 3. $\Omega^c \mid A \succcurlyeq C^c \mid B \succcurlyeq \emptyset^c \mid A$;

Axiom 4a. If $X_1 \mid A \succcurlyeq Y_1 \mid B$ and $X_2 \mid A \succcurlyeq Y_2 \mid B$ then

$$X_1 \mid A + X_2 \mid A \succcurlyeq Y_1 \mid B + Y_2 \mid B.$$

Axiom 4b. If $X_1 \mid A \preccurlyeq Y_1 \mid B$ and $X_1 \mid A + X_2 \mid A \succcurlyeq Y_1 \mid B + Y_2 \mid B$ then

$$X_2 \mid A \succcurlyeq Y_2 \mid B.$$

Axiom 5. If $A \subseteq B$ then

$$X \mid A \succcurlyeq Y \mid A \quad iff \quad X \cdot A^c \mid B \succcurlyeq Y \cdot A^c \mid B;$$

Axiom 6. (Archimedean). If $X \mid A \succ Y \mid B$ then for every Z in $\mathcal{F}^*$ there is a positive integer n such that

$$nX \mid A \succcurlyeq nY \mid B + Z \mid B.$$

The axioms are simple in character and their relation to the axioms of Suppes and Zanotti (1976) is apparent. The first three axioms are very similar. Axiom 4, the axiom of addition, must be relativized to the restricted set. Notice that we have a different restriction on the two sides of the inequality. We have been unable to show whether or not it is possible to replace the two parts of Axiom 4 by the following weaker and more natural axiom. If $X_2 \mid A \sim Y_2 \mid B$, then $X_1 \mid A \succcurlyeq Y_1 \mid B$ iff $X_1 \mid A + X_2 \mid A \succcurlyeq Y_1 \mid B + Y_2 \mid B$.

The really new axiom is Axiom 5. In terms of events and numerical probability, this axiom corresponds to the following: If $A \subseteq B$, *then* $P(C \mid A) \succcurlyeq P(D \mid A)$ *iff* $P(C \cap A \mid B) \succcurlyeq P(D \cap A \mid B)$. Note that in the axiom itself, function multiplication replaces intersection of events. (Closure of $\mathcal{F}^*$ under function multiplication is easily proved.) This axiom does not seem to have previously been used in the literature. Axiom 6 is the familiar and necessary Archimedean axiom.

3. REPRESENTATION THEOREM

We now state and prove the main theorem of this paper. In the theorem we refer to a strictly agreeing expectation function on $\mathcal{RF}^*(\Omega)$. From standard probability theory and conditional expected utility theory, it is evident that the properties of this expectation should be the following for $A, B \succ \emptyset$.

12

(i) $E(X \mid A) \succcurlyeq E(Y \mid B)$ iff $X \mid A \succcurlyeq Y \mid B$,

(ii) $E(X \mid A + Y \mid A) = E(X \mid A) + E(Y \mid A)$,

(iii) $E(X \cdot A^c \mid B) = E(X \mid A)E(A^c \mid B)$ if $A \subseteq B$,

(iv) $E(\emptyset^c \mid A) = 0$ and $E(\Omega^c \mid A) = 1$.

Using primarily (iii), it is then easy to prove the following property, which occurs in the earlier axiomatic literature mentioned above:

$$E(X \mid A \cup Y \mid B) = E(X \mid A)E(A^c \mid A \cup B) + E(Y \mid B)E(B^c \mid A \cup B).$$

for $A \cap B = \emptyset$

THEOREM *Let Ω be a nonempty set, let $\mathcal{F}$ be an algebra of sets on Ω, and let $\succcurlyeq$ be a binary relation on $\mathcal{F} \times \mathcal{F}$. Then a necessary and sufficient condition that there is a strictly agreeing conditional probability measure on $\mathcal{F} \times \mathcal{F}$ is that there is an extension $\succcurlyeq^*$ of $\succcurlyeq$ from $\mathcal{F} \times \mathcal{F}$ to $\mathcal{RF}^*(\Omega)$ such that the structure $(\Omega, \mathcal{RF}^*(\Omega), \succcurlyeq^*)$ is a partial qualitative expectation structure. Moreover, if $(\Omega, \mathcal{RF}^*(\Omega), \succcurlyeq^*)$ is a partial qualitative expectation structure, then there is a unique strictly agreeing expectation function on $\mathcal{RF}^*(\Omega)$ and this expectation generates a unique strictly agreeing conditional probability measure on $\mathcal{F} \times \mathcal{F}$.*

Proof. For every $X \mid A$, with $A \succ \emptyset$, we define the set

$$S(X \mid A) = \left\{ \frac{m}{n} : m\Omega^c \mid A \succcurlyeq nX \mid A \right\}$$

(We note that it is easy to prove from the axioms that $\Omega^c \sim \Omega^c \mid A$, and thus for general purposes we can write: $m\Omega^c \succcurlyeq nX \mid A$.) Given this definition, on the basis of the reduction by Suppes and Zanotti (1976) of Axioms 1–4 and 6 to a known necessary and sufficient condition for extensive measurement (Krantz *et al.*, 1971, chap. 3), we know first that the greatest lower bound of $S(X \mid A)$ exists, and following the proof in Krantz *et al.* we use this to define the expectation of X given A:

(1) $$E(X \mid A) = \text{g.l.b.} \left\{ \frac{m}{n} : m\Omega^c \succcurlyeq nX \mid A \right\}.$$

It then follows from these earlier results that the function E (for fixed A) is unique and:

(2) $$E(X \mid A) \geqslant E(Y \mid A) \quad \text{iff} \quad X \mid A \succcurlyeq Y \mid A.$$

(3) $$E(X \mid A + Y \mid A) = E(X \mid A) + E(Y \mid A).$$

(4) $$E(\emptyset^c \mid A) = 0 \quad \text{and} \quad E(\Omega^c \mid A) = 1.$$

The crucial step is now to extend the results to the relation between given events A and B.

13

We first prove the preservation of order by the expectation function. For the first half of the proof, assume

(5) $$X \mid A \succcurlyeq Y \mid B,$$

and suppose, on the contrary, that

(6) $$E(Y \mid B) > E(X \mid A).$$

Then there must exist natural numbers m and n such that

(7) $$E(Y \mid B) > \frac{m}{n} > E(X \mid A),$$

and so from the definition of the function E, we have

(8) $$m\Omega^c \prec nY \mid B,$$

and

(9) $$m\Omega^c \succcurlyeq nX \mid A,$$

whence

(10) $$nY \mid B \succ nX \mid A,$$

but from (5) and Axiom 4a we have by a simple induction

(11) $$nX \mid A \succcurlyeq nY \mid B,$$

which contradicts (10), and thus the supposition (6) is false.

Assume now

(12) $$E(X \mid A) \geqslant E(Y \mid B),$$

and suppose

(13) $$Y \mid B \succ X \mid A.$$

Now if $E(X \mid A) > E(Y \mid B)$, by the kind of argument just given we can show at once that

(14) $$X \mid A \succ Y \mid B,$$

which contradicts (13). On the other hand, if

(15) $$E(X \mid A) = E(Y \mid B),$$

then we can argue as follows. By virtue of (13) and Axiom 6, there is an n such that

(16) $$nY \mid B \succcurlyeq (n + 1)X \mid A,$$

whence by the earlier argument

(17) $$E(nY \mid B) \geqslant E((n+1)X \mid A),$$

and by (3)

(18) $$nE(Y \mid B) \geqslant (n+1)E(X \mid A),$$

and so by (15) and (18)

(19) $$E(Y \mid B) \leqslant 0,$$

but from (2)–(4) it follows easily that

(20) $$E(Y \mid B) \geqslant 0,$$

whence

(21) $$E(Y \mid B) = 0,$$

but then, using again (2)–(4), we obtain

(22) $$Y \mid B \sim \emptyset^c \mid B,$$

and by virtue of Axiom 3

(23) $$X \mid A \succcurlyeq \emptyset^c \mid B,$$

whence from (22) and (23) by transitivity

(24) $$X \mid A \succcurlyeq Y \mid B,$$

contradicting (13). We have thus now shown that

(25) $$E(X \mid A) \geqslant E(Y \mid B) \quad \text{iff} \quad X \mid A \succcurlyeq Y \mid B.$$

Finally, we need to prove that for $A \succ 0$ and $A \subseteq B$

(26) $$E(X \cdot A^c \mid B) = E(X \mid A)E(A^c \mid B).$$

We first note that by putting $m\Omega^c$ for X and nX for Y in Axiom 5, we obtain

(27) $$m\Omega^c \succcurlyeq nX \mid A \quad \text{iff} \quad mA^c \mid B \succcurlyeq nX \cdot A^c \mid B.$$

It follows directly from (27) that

(28) $$\left\{ \frac{m}{n} : m\Omega^c \succcurlyeq nX \mid A \right\} = \left\{ \frac{m}{n} : mA^c \mid B \succcurlyeq nX \cdot A^c \mid B \right\},$$

whence their greatest lower bounds are the same, and we have

(29) $$E(X \mid A) = E'_{A^c \mid B}(X \cdot A^c \mid B),$$

where E' is the measurement function that has $A^c \mid B$ as a unit, that is,

$$E'_{A^c \mid B}(A^c \mid B) = 1.$$

15

As is familiar in the theory of extensive measurement, there exists a positive real number c such that for every X

(30) $$cE'_{A^c \mid B}(X \cdot A^c \mid B) = E(X \cdot A^c \mid B).$$

Now by (29) and taking $X = \Omega^c$

$$cE(\Omega^c \mid A) = E(\Omega^c \cdot A^c \mid B),$$

but $E(\Omega^c \mid A) = 1$, so

(31) $$c = E(\Omega^c \cdot A^c \mid B) = E(A^c \mid B).$$

Combining (29), (30), and (31) we obtain (26) as desired.

The uniqueness of the expectation function follows from (4) and the earlier results (Suppes and Zanotti, 1976) about unconditional probability.

For $A \succ \varnothing$, we then define for every B in $\mathcal{F}$,

$$P(B \mid A) = E(B^c \mid A),$$

and it is trivial to show that the function P is a conditional probability measure on $\mathcal{F}$, which establishes the sufficiency of the axioms. The necessity of each of the axioms is easily checked.

REFERENCES

Aczél, J.: 1961, Über die Begründung der Additions- und Multiplikationsformeln von bedingten Wahrscheinlichkeiten. Magyar Tudományos Akadémia Matematikai Kutató Int. Közleményei, 6, 110–122.

Aczél, J.: 1966, Lectures on functional equations and their applications. New York: Academic Press.

Domotor, Z: 1969, Probabilistic relational structures and their applications (Tech. Rep. 144). Stanford, Calif.: Stanford University, Institute for Mathematical Studies in the Social Sciences.

Koopman, B. O.: 1940a, The axioms and algebra of intuitive probability. Ann. Math., 41. 269–292.

Koopman, B. O.: 1940b, The bases of probability. Bull. Amer. Math. Soc., 46, 763–774.

Krantz. D. H., Luce, R. D., Suppes, P., Tversky, A.: 1971, Foundations of Measurement (Vol. I). New York: Academic Press.

Luce, R. D.: 1968, On the numerical representation of qualitative conditional probability. Ann. Math. Statis., 39, 481–491.

Roberts, F. S., Luce, R. D.: 1968, Axiomatic thermodynamics and extensive measurement. Synthese, 18, 311–326.

Scott, D.: 1964, Measurement structures and linear inequalities. Journal of Mathematical Psychology, 1. 233–247.

Suppes, P.: 1973, New foundations of objective probability: Axioms for propensities. In P. Suppes, L. Henkin, G. C. Moisil, A. Joja (Eds.), Logic, Methodology, and Philosophy of Science IV: Proceedings of the Fourth International Congress for Logic, Methodology and Philosophy of Science, Bucharest, 1971, pp. 515–529. Amsterdam: North-Holland.

Suppes, P., Zanotti, M.: 1976, Necessary and sufficient conditions for existence of a unique measure strictly agreeing with a qualitative probability ordering. Journal of Philosophical Logic, 5, 431–438. [chap. 1, this volume].

3

On using random relations to generate upper and lower probabilities

For a variety of reasons there has been considerable interest in upper and lower probabilities as a generalization of ordinary probability. Perhaps the most evident way to motivate this generalization is to think of the upper and lower probabilities of an event as expressing bounds on the probability of the event. The most interesting case conceptually is the assignment of a lower probability of zero and an upper probability of one to express maximum ignorance.

Simplification of standard probability spaces is given by random variables that map one space into another and usually simpler space. For example, if we flip a coin a hundred times, the sample space describing the possible outcome of each flip consists of 2^{100} points, but by using the random variable that simply counts the number of heads in each sequence of a hundred flips we can construct a new space that contains only 101 points. Moreover, the random variable generates in a direct fashion the appropriate probability measure on the new space.

What we set forth in this paper is a similar method for generating upper and lower probabilities by means of random relations. The generalization is a natural one; we simply pass from functions to relations, and the multivalued character of the relations leads in an obvious way to upper and lower probabilities.

The generalization from random variables to random relations also provides a method for introducing a distinction between indeterminacy and uncertainty that we believe is new in the literature. Both of these concepts are defined in a purely set-theoretical way and thus do not depend, as they often do in informal discussions, on explicit probability considerations. Random variables, it should be noted, possess uncertainty but not indeterminacy. In this sense, the concept of indeterminacy is a generalization that goes strictly beyond ordinary probability theory, and thus provides a means of expressing the intuitions of

Reprinted from *Synthese* 36 (1977), 427–440.

those philosophers who are not satisfied with a purely probabilistic notion of indeterminacy.

Section 1 is devoted to set-theoretical concepts. The initial developments are standard and follow familiar ideas in the set-theoretical approach to relations and operations on relations (see Suppes, 1972, chap. 3). On the other hand, to achieve the appropriate set-theoretical version of a lower probability, we introduce the notion of lower image, which is not standard in the set-theoretical literature and is, as far as we know, introduced as an explicit concept for the first time in this paper. It is interesting to note that the upper and lower images introduced in Section 1 have set-theoretical properties that are essentially complete analogues of the properties of upper and lower probabilities.

In Section 2 we turn to upper and lower probabilities and state a fundamental representation theorem showing how those upper and lower probabilities generated by random relations can always be intrinsically characterized in terms of capacities of infinite order. Here and in other parts of the paper, basic theorems are stated but no proofs are given.

There is no single concept of upper or lower conditional probability, in the sense that there is a single conditional probability concept corresponding to a given probability measure. We consider in Section 2 the two most obvious candidates and show that the property of being a capacity of infinite order is preserved by them. Finally, we examine informally some of the difficulties of using Bayes' theorem in a direct way in this generalized setup. These difficulties show that, in spite of the intuitive appeal of the use of upper and lower probabilities to express ignorance, there are subtle problems in developing an intuitively reasonable theory of inference.

1. SET-THEORETICAL UPPER AND LOWER IMAGES

Let X and Y be two nonempty sets. Then the set $\mathbf{R}(X, Y)$ is the set of all (binary) relations $R \subseteq X \times Y$. We shall also occasionally refer to such a relation R as a multivalued mapping from X into Y. It is obvious that $\mathbf{R}(X, Y)$ is a Boolean algebra under the operations of intersection, union and complementation.

The *domain* of a relation R is defined as

$$(1) \qquad \mathcal{D}(R) = \{x: (\exists y)(xRy)\},$$

and the notion of *range* is defined similarly,

$$(2) \qquad \mathcal{R}(R) = \{y: (\exists x)(xRy)\}.$$

The domain function $\mathcal{D}$ may also be thought of as a mapping from $\mathbf{R}(X, Y)$ to the power set, $\mathcal{P}(X)$, of X, and the range function as a mapping from $\mathbf{R}(X, Y)$ to $\mathcal{P}(Y)$.

Because of the symmetry in the domain and range mappings, we list explicitly only the properties of the domain mapping:

(3) $\mathcal{D}(\emptyset) = \emptyset$,   where $\emptyset$ is the empty set, which is also the empty relation,

(4) $\mathcal{D}(U) = X$,   where $U = X \times Y$ is the universal relation,

(5) $\mathcal{D}(R_1 \cup R_2) = \mathcal{D}(R_1) \cup \mathcal{D}(R_2)$,   for $R_1, R_2 \in \mathbf{R}(X, Y)$,

(6) $\mathcal{D}(R_1 \cap R_2) \subseteq \mathcal{D}(R_1) \cap \mathcal{D}(R_2)$,

(7) $\mathcal{D}(R_1) \sim \mathcal{D}(R_2) \subseteq \mathcal{D}(R_1 \sim R_2)$,   where $\sim$ is set difference.

For several purposes it is convenient to have a restricted form of complementation: For $R \in \mathbf{R}(X, Y)$ the complement $\neg R$ is with respect to $X \times Y$, i.e.,

(8) $$\neg R = (X \times Y) \sim R,$$

the complement of $A \subseteq X$ is $X \sim A$, and the complement of $B \subseteq Y$ is $Y \sim B$. Thus $\neg \mathcal{D}(R) = X \sim \mathcal{D}(R)$. The point to note is that unrestricted complementation of sets is of no interest in the present context, i.e., it is of no interest to have the complementation of $R \in \mathbf{R}(X, Y)$ and $\mathcal{D}(R)$ relative to the same universe.

We next turn to some familiar operations on relations, or on relations and sets. The *converse* or *inverse* of a relation is defined as

(9) $$\check{R} = \{(y, x) : xRy\}.$$

This notion is, of course, the relational generalization of function inverse. Familiar properties for R, R_1 and R_2 in $\mathbf{R}(X, Y)$ are these:

(10) $$\check{\check{R}} = R,$$

(11) $$\overbrace{R_1 \cap R_2} = \check{R}_1 \cap \check{R}_2,$$

(12) $$\overbrace{R_1 \cup R_2} = \check{R}_1 \cup \check{R}_2,$$

(13) $$\overbrace{R_1 \sim R_2} = \check{R}_1 \sim \check{R}_2.$$

The notion of a relation's domain being restricted to a given set is defined as

(14) $$R \mid A = R \cap (A \times \mathcal{R}(R)).$$

Thus if $R = \{(1, a), (1, b), (2, c), (3, b)\}$ and $A = \{2, 3\}$, then $R \mid A = \{(2, c), (3, b)\}$. Elementary properties of the restriction operation are these:

(15)   If $A \subseteq B$ then $R \mid A \subseteq R \mid B$,

(16) $R \mid (A \cap B) = (R \mid A) \cap (R \mid B)$,

(17) $R \mid (A \cup B) = (R \mid A) \cup (R \mid B)$,

(18) $R \mid (A \sim B) = (R \mid A) \sim (R \mid B)$.

19

We next turn to two concepts that are especially important for subsequent developments. The first, R"A, is ordinarily called the *image* of A under the relation R, but, for reasons that will soon be made clear, we shall call it the *upper image* of A under R. The definition is simple in terms of restriction and range.

$$(19) \qquad R\text{"}A = \mathcal{R}(R \mid A),$$

but more suggestive is the equivalence

$$(20) \qquad y \in R\text{"}A \quad \text{iff} \quad (\exists x)(xRy \text{ and } x \in A).$$

Now let us define, for immediate comparison, the less standard concept of *lower image*, introduced in analogy to the relation between upper and lower probabilities:

$$(21) \qquad P_*(A) = 1 - P^*(\neg A).$$

Thus, we have

$$(22) \qquad R_{\prime\prime}A = \neg(R\text{"}\neg A).$$

The 'outside' complementation of (22) is with respect to Y, and the 'inside' one with respect to X. In order to have, again in analogy to the case of upper and lower probabilities, the inequality corresponding to

$$(23) \qquad P_*(A) \leqslant P^*(A),$$

we need for the range of R to be Y, and in the case of the inverse image, the range of $\breve{R}$ to be X.

$$(24) \qquad \begin{array}{lll} \text{If } \mathcal{R}(R) = Y & \text{then} & R_{\prime\prime}A \subseteq R\text{"}A. \\ \text{If } \mathcal{R}(\breve{R}) = X & \text{then} & \breve{R}_{\prime\prime}B \subseteq \breve{R}\text{"}B. \end{array}$$

This restriction is a natural one, for it corresponds to a multivalued mapping having all of X as its domain, a point that is expanded on below.

The familiar subadditive and superadditive properties of upper and lower probabilities are expressed in the inequalities: For $A \cap B = \emptyset$,

$$(25) \quad P_*(A) + P_*(B) \leqslant P_*(A \cup B) \leqslant P^*(A \cup B) \leqslant P^*(A) + P^*(B).$$

As the relational analogue we have:

$$(26) \qquad (R_{\prime\prime}A) \cup (R_{\prime\prime}B) \subseteq R_{\prime\prime}(A \cup B),$$

$$(27) \qquad R\text{"}(A \cup B) = (R\text{"}A) \cup (R\text{"}B),$$

20

and (26) and (27) are not restricted to $A \cap B = \emptyset$. Some other properties of the upper and lower images of a set are the following:

$$(28) \qquad R_{\cdot\cdot}(A \cap B) = (R_{\cdot\cdot}A) \cap (R_{\cdot\cdot}B),$$

$$(29) \qquad R^{\cdot\cdot}(A \cap B) \subseteq (R^{\cdot\cdot}A) \cap (R^{\cdot\cdot}B),$$

$$(30) \qquad \text{If } A \subseteq B \text{ then } R^{\cdot\cdot}A \subseteq R^{\cdot\cdot}B,$$

$$(31) \qquad \text{If } A \subseteq B \text{ then } R_{\cdot\cdot}A \subseteq R_{\cdot\cdot}B,$$

$$(32) \qquad R_{\cdot\cdot}\emptyset = R^{\cdot\cdot}\emptyset = \emptyset,$$

$$(33) \qquad R_{\cdot\cdot}X = R^{\cdot\cdot}X = \mathcal{R}(R).$$

Note that in (33) $\mathcal{R}(R)$ plays the role of the universe in the image sample space. On the basis of (28) the lower image is a homomorphism with respect to the intersection of sets, and on the basis of (27) the upper image is such a mapping with respect to the union of sets.

We now turn to relations between Boolean algebras on X and Y. Given $R \in \mathbf{R}(X, Y)$ and a Boolean algebra $\mathcal{B}$ of subsets of Y, the class

$$(34) \qquad \mathcal{C}_* = \{A : A \subseteq X \ \& \ (\exists B)(B \in \mathcal{B} \ \& \ \check{R}_{\cdot\cdot}B = A)\}$$

is a π-system of subsets of X, i.e., it is closed under intersection, and the class

$$(35) \qquad \mathcal{C}^* = \{A : A \subseteq X \ \& \ (\exists B)(B \in \mathcal{B} \ \& \ \check{R}^{\cdot\cdot}B = A)\}$$

is a family of subsets of X closed under union. The classes $\mathcal{C}_*$ and $\mathcal{C}^*$ are said to be *induced* from $\mathcal{B}$ by R. If R is a function from X to Y, then $\mathcal{C}_*$ and $\mathcal{C}^*$ are Boolean algebras and $\mathcal{C}_* = \mathcal{C}^*$.

It is clear that $\mathcal{C}_*$ and $\mathcal{C}^*$ each generate Boolean algebras on X, by adding closure under complementation. We have the following:

THEOREM 1 *Let $\mathcal{B}(\mathcal{C}_*)$ and $\mathcal{B}(\mathcal{C}^*)$ be the Boolean algebras on X generated by $\mathcal{C}_*$ and $\mathcal{C}^*$, respectively. Then*

$$(36) \qquad \mathcal{B}(\mathcal{C}_*) = \mathcal{B}(\mathcal{C}^*).$$

We next introduce the concept of a measurable relation, which is a natural generalization of the standard concept of measurable function. Recall first that a measurable space $(X, \mathcal{B})$ consists of a nonempty set X and a Boolean algebra of subsets of X. Given two measurable spaces $(X, \mathcal{B}_1)$ and $(Y, \mathcal{B}_2)$, a relation $R \in \mathbf{R}(X, Y)$ is said to be $(\mathcal{B}_1, \mathcal{B}_2)$-*measurable* if $\check{R}_{\cdot\cdot}\mathcal{B}_2$ and $\check{R}^{\cdot\cdot}\mathcal{B}_2$

are contained in $\mathcal{B}_1$. Here extension of the upper and lower image notation to families of sets is obvious; e.g.,

(37) $$\check{R}_{..}\mathcal{B}_2 = \{A : A \subseteq X \;\&\; (\exists B)(B \in \mathcal{B}_2 \;\&\; \check{R}_{..}B = A)\}.$$

We have then the following theorem.

THEOREM 2 *In order that $R \in \mathbf{R}(X, Y)$ be $(\mathcal{B}_1, \mathcal{B}_2)$-measurable it suffices that either $\check{R}_{..}\mathcal{B}_2 \subseteq \mathcal{B}_1$ or $R``\mathcal{B}_2 \subseteq \mathcal{B}_1$.*

We conclude this section by defining the concepts of uncertainty and indeterminacy for random relations. The intuitive content of these two notions is discussed near the beginning of the next section.

Given R_1 and R_2 in $\mathbf{R}(X, Y)$, we say that R_1 is *less certain* than R_2 iff the domain of R_2 is a proper subset of the domain of R_1, i.e., $\mathcal{D}(R_2) \subset \mathcal{D}(R_1)$, and we say that R_1 is *less determined* than R_2 iff

(i) $R_2``\{x\} \subseteq R_1``\{x\}$ for all $x \in \mathcal{D}(R_1 \cap R_2)$;
(ii) $R_2``\{x\} \subset R_1``\{x\}$ for some $x \in \mathcal{D}(R_1 \cap R_2)$.

We illustrate these fundamental comparative ideas of uncertainty and indeterminacy by a simple example that is just barely complex enough to provide a basis for meaningful distinctions. Suppose we have two coins, one new and one badly worn. We flip them together and record in our sample space representation the outcome for the new coin followed by the outcome for the worn coin. Thus

$$X = \{hh, ht, th, tt\},$$

and the outcome ht, for example, means that the new coin came up 'heads' and the worn coin 'tails'. Suppose next that we are only interested in the number of heads. Thus

$$Y = \{0, 1, 2\}.$$

Now suppose, and this is the crucial assumption, that we can easily misread the face of the worn coin, but do not make any mistakes about the face of the new coin. This essential aspect of the situation is represented by the relation R (or multivalued map) from X to Y. $R``\{hh\} = \{1, 2\}$, because the second $h$ could be read as $t$, $R``\{ht\} = \{1, 2\}$, $R``\{th\} = \{0, 1\}$ and $R``\{tt\} = \{0, 1\}$. Now let us compare with R the standard random variable, say, S, that counts correctly the number of heads in any outcome. The relation S is then, of course, a function: $S``\{hh\} = 2$, etc. Note now that $\mathcal{D}(R) = \mathcal{D}(S) = X$, and $\mathcal{R}(R) = \mathcal{R}(S) = Y$. Thus according to the definitions given R and S are equivalent in uncertainty, but R is less determined than S, for

22

(i) $S"\{x\} \subseteq R"\{x\}$ for all x in X,

(ii) $S"\{hh\} \subset R"\{hh\}$.

On the other hand, suppose we know that the observed number of heads is 0 or 1. Let $A = \{0, 1\}$. Then we generate two new relations

$$R_1 = R \cap (\mathcal{D}(R) \times A),$$
$$S_1 = S \cap (\mathcal{D}(S) \times A),$$

and we see at once that

$$\mathcal{R}(R_1) = \mathcal{R}(S_1) = A,$$

and thus we can assert: R and S are less certain than S_1. At the same time, S and S_1 are equivalent in indeterminacy, and R is less determined than R_1.

This trivial example of coin flipping can intuitively illustrate several important conceptual points about the concepts of uncertainty and indeterminacy we have introduced.

(i) The reduction of uncertainty in going from the relation R or S to S_1, corresponds to conditionalizing on a known event in the ordinary probability theory of events. Moreover, there is a simple set-theoretical way of showing this restriction in terms of the notation introduced in (14), namely,

$$\check{R}_1 = \check{R} \mid A,$$
$$\check{S}_1 = \check{S} \mid A.$$

(ii) For a strict Bayesian there is no indeterminacy, for he would postulate a prior probability of misreading the worn coin, and thereby obtain a standard random variable, i.e., function, mapping X into Y. The concept of indeterminacy is a concept for those who hold that not all sources of error, lack of certain knowledge, etc., are to be covered by a probability distribution, but may be expressed in other ways, in particular, by random relations as generalizations of random variables, and by the resulting concepts of upper and lower probabilities. Without trying to give a conclusive argument, we feel there are good grounds for believing that maximum indeterminacy as defined here, i.e., $P_*(A) = 0$ and $P^*(A) = 1$, rather than something like a uniform prior distribution, is a more natural way to express maximum ignorance about the true state of affairs in a wide variety of situations.

(iii) Two random variables, i.e., measurable functions, from X to Y – arbitrary X and Y, not just the examples given above – are equivalent in determinacy and are in fact maximally determinate. Thus, if f is a measurable function from X to Y, there is no relation R from X to Y such that f is less determined than R. This result is another way of stating that only random relations that are strict generalizations of random variables possess indeterminacy.

In one respect, our trivial example has one unfortunate feature. The most appealing applications of the concept of indeterminacy permit continual reduction of the indeterminacy by experimentation. This feature is not salient in the above example, so we think it useful to introduce a second example, also trivial, but possessing this reduction feature.

Suppose someone is about to flip a coin that we have not seen. The context is such that we have three different hypotheses about the coin: H_1 – it has heads on one side and tails on the other; H_2 – both sides are heads; H_3 – both sides are tails. Let $\mathcal{H} = \{H_1, H_2, H_3\}$. The sample space of possible observations is $\Omega = \{h, t\}$. It is not entirely obvious how to construct the appropriate space X for a problem of this kind. The usual statistical approach is to take the product space $\Omega \times \mathcal{H}$, but to express the appropriate indeterminacy about the hypotheses. This does not work out. What we use instead is the space $\Omega^{\mathcal{H}}$ of functions from $\mathcal{H}$ to Ω, but we delete from this space the functions ruled out as impossible by the hypotheses, and thus we have left the two functions: $f_1(H_1) = h$, $f_1(H_2) = h$, $f_1(H_3) = t$ and $f_2(H_1) = t$, $f_2(H_2) = h$, $f_2(H_3) = t$, so that $X = \{f_1, f_2\}$. We now define on X three random variables R_1, R_2, and R_3 with R_i corresponding to H_i. The random variable R_i counts the number of heads in each point of X according to hypothesis H_i. We then take as our random relation

$$R = R_1 \cup R_2 \cup R_3.$$

It is then easy to check that $R``\{f_1\} = R``\{f_2\} = \{0, 1\}$, where $Y = \{0, 1\}$, and thus R has maximal indeterminacy, which expresses our maximal ignorance about the true hypothesis.

2. UPPER AND LOWER PROBABILITIES

We show in this section how, given a probability space, a random relation generates an upper and lower probability on the image space. Here and in what follows we use only finite additivity, and also in our earlier definition of measurability we assume only Boolean algebras of sets, not σ-algebras closed under denumerable unions. The extension of measurability and of the probability space to countable closure is direct and requires only minor technical changes in our formulation. To signal that probability is now being considered explicitly, we print random variables and random relations in boldface from this point on.

Given a measurable space $(Y, \mathcal{B}_2)$, a probability space $\mathcal{X} = (X, \mathcal{B}_1, P)$, and a $(\mathcal{B}_1, \mathcal{B}_2)$-measurable relation $\boldsymbol{R} \in \mathbf{R}(X, Y)$, we define for $A \in \mathcal{B}_2$

(38)
$$P_*(A) = P(\check{\boldsymbol{R}}_{..}A)$$
$$P^*(A) = P(\check{\boldsymbol{R}}``A).$$

24

We call the pair (P_*, P^*) a *Dempsterian* functional (generated by $\mathcal{X}$ and $\boldsymbol{R}$) after Dempster (1967).

Our first trivial example to illustrate indeterminacy may also be used to illustrate the definitions embodied in (38). Let both the new and the worn coin be fair, then the probability of each atom in X is .25. In the case of $Y = \{0, 1, 2\}$, we have then, e.g.,

$$P^*(\{2\}) = P(\check{\boldsymbol{R}}^{\text{``}}\{2\})$$
$$= P(\{hh, ht\})$$
$$= .5,$$

and

$$P_*(\{2\}) = P(\check{\boldsymbol{R}}_{\iota\iota}\{2\})$$
$$= P(\neg(\check{\boldsymbol{R}}^{\text{``}}\neg\{2\}))$$
$$= 1 - P(\check{\boldsymbol{R}}^{\text{``}}\neg\{2\})$$
$$= 1 - P(\check{\boldsymbol{R}}^{\text{``}}\{0, 1\})$$
$$= 1 - P(\{hh, ht, th, tt\})$$
$$= 0.$$

By similar analysis, $P_*(\{0\}) = P_*(\{1\}) = 0$, and $P^*(\{0\}) = .5$, $P^*(\{1\}) = 1$.

More generally, if $(Y, \mathcal{B})$ is a measurable space, a pair (P_*, P^*) of real-valued functions defined on $\mathcal{B}$ is said to be an *upper–lower functional* iff $P_*(Y) = P^*(Y) = 1$, for every A in $\mathcal{B}$, $P_*(A) \geqslant 0$, and inequality (25) holds for disjoint sets A and B in $\mathcal{B}$. It is obvious that every Dempsterian functional is an upper–lower functional. The converse is, of course, false.

An upper–lower functional (P_*, P^*) on a measurable space $(Y, \mathcal{B})$ is said to be a *capacity of order n* iff the following conditions are satisfied for all $A, A_1, \ldots, A_n$ in $\mathcal{B}$:

(i) $P_*(A) - \sum_i P_*(A \cap A_i) + \sum_{i,j} P_*(A \cap A_i \cap A_j) + \cdots + (-1)^n P_*(A \cap A_1 \cap \cdots \cap A_n) \geqslant 0$,

(ii) $P^*(A) = 1 - P_*(\neg A)$.

Obviously, if (P_*, P^*) is a capacity of order n, then it is a capacity of order $m \leqslant n$. In addition, we say that (P_*, P^*) is a capacity of *infinite order* if it is a capacity of order n for all $n \geqslant 1$. The concept of capacity is thoroughly studied by Choquet (1955). We have two fundamental theorems relating Dempsterian functionals and capacities of infinite order.

THEOREM 3 *Given a measurable space* $(Y, \mathcal{B}_2)$, *a probability space* $\mathcal{X} = (X, \mathcal{B}_1, P)$, *and a* $(\mathcal{B}_1, \mathcal{B}_2)$-*measurable relation* $\boldsymbol{R} \in R(X, Y)$, *then the Dempsterian functional* (P_*, P^*) *generated by* $\mathcal{X}$ *and* $\boldsymbol{R}$ *is a capacity of infinite order.*

THEOREM 4 *Given a measurable space* $(Y, \mathcal{B}_2)$ *and an upper–lower functional* (P_*, P^*) *that is a capacity of infinite order on the space, then there is a probability space* $\mathcal{X} = (X, \mathcal{B}_1, P)$ *and a random relation* $\mathbf{R} \in \mathbf{R}(X, Y)$ *such that* (P_*, P^*) *is a Dempsterian functional generated by* $\mathcal{X}$ *and* $\mathbf{R}$.

These two theorems taken together provide a fundamental representation theorem for upper–lower probability functionals (P_*, P^*). In order for such a functional to have been generated from an underlying probability space by a random relation it is necessary and sufficient that it be a capacity of infinite order.

It is worth noting that significant classes of upper and lower probabilities are not capacities of infinite order. For instance, let $\mathcal{P}$ be a nonempty set of probability measures on a measurable space $(X, \mathcal{B})$. Define for each A in $\mathcal{B}$

$$P_*(A) = \inf_{P \in \mathcal{P}} P(A)$$

$$P^*(A) = \sup_{P \in \mathcal{P}} P(A),$$

then in general the upper–lower functional (P_*, P^*) will not be a capacity of infinite order, and thus cannot be generated by a random relation on a probability space.

As a second example, the upper and lower probabilities that are constructed in the theory of approximate measurement developed in Suppes (1974) are in general not even capacities of order two. Thus the upper and lower probabilities arising from approximations in measurement are about as far from being capacities of infinite order as it is possible to be.

Conditionalization. We now turn to the upper and lower analogues of conditional probability. The first and perhaps most fundamental point to note is that there is not one single concept corresponding to ordinary conditional probability. Following Dempster (1967), we define the *Dempsterian upper–lower conditional probabilities* as follows, for $P^*(A) \neq 0$,

(39)
$$\begin{cases} P^*(B \mid A) = P^*(B \cap A)/P^*(A) \\ P_*(B \mid A) = 1 - P^*(\neg B \mid A). \end{cases}$$

and the *geometric upper–lower conditional probabilities* by

(40)
$$\begin{cases} P_*(B \mid A) = P_*(B \cap A)/P_*(A) \\ P^*(B \mid A) = 1 - P_*(\neg B \mid A). \end{cases}$$

It is easy to give examples that show that Dempsterian and geometric upper–lower conditional probabilities are not the same. The name *geometric* is used because of properties shared with the classical concept of geometric probability.

Both of these methods of conditionalizing preserve the Dempsterian or capacity property of an upper–lower functional. More explicitly, we have the following theorem.

26

THEOREM 5 *Given a measurable space* $(X, \mathcal{B})$ *and an upper–lower functional* (P_*, P^*) *that is a capacity of infinite order on the space, then either the Dempsterian or geometric conditionals, as defined by* (39) *and* (40) *are also capacities of infinite order on the space.*

These two different conditionals can be generated by natural random relations – the existence of some such random relations is guaranteed by Theorems 4 and 5.

We shall not set out the details here, but it can be shown that the random relation generating the geometrical upper–lower conditional probabilities can reduce uncertainty but not indeterminacy. In contrast, the random relation generating the Dempsterian conditional can reduce both uncertainty and indeterminacy.

Given these two different conditionals, it is natural to ask which one should be used for inference. Dempster (1967, 1968) has developed a theory of inference around his concept, but it has been sharply criticized and above all does not seem to be based on intuitively appealing principles that have a clear and straightforward statement.

We are not prepared to offer an alternative in the present framework, but we want to conclude by pointing out why a simple generalization of Bayes' theorem will not work for upper and lower probabilities, and why the theory of inference for such probabilities is a good deal more difficult and subtle than it might seem to be upon casual inspection. For reference we state Bayes' theorem in both an upper and a lower form, and we suppose a finite set of hypotheses $H_1, \ldots, H_n$, and evidence E as events–a more complicated formulation is not needed in the present context.

(41) $$P_*(H_i \mid E) = P_*(E \mid H_i) P_*(H_i) / P_*(E),$$

(42) $$P^*(H_i \mid E) = P^*(E \mid H_i) P^*(H_i) / P^*(E).$$

First, in the case of either (41) or (42) we ordinarily cannot compute the denominator, so we have to retreat to a proportionality statement, which in itself is not too serious.

Second, and far more serious, even if we ignore the denominator, given the prior $P_*(H_i)$ and $P^*(H_i)$ and the likelihood, which in many cases is a probability, $P_*(E \mid H_i) = P^*(E \mid H_i)$, we cannot compute the conditional upper or lower probabilities for other than individual hypotheses. For example, given $P_*(H_1 \mid E)$ and $P_*(H_2 \mid E)$, we cannot compute $P_*(\{H_1, H_2\} \mid E)$, for all we know within the framework of (41) is that the lower conditional probability is superadditive, and thus satisfies inequality (25) rather than an equality. Similar remarks apply to the upper conditional probability.

Third, let us restrict ourselves drastically to the posterior for individual hypotheses, and the conceptually interesting case of maximum ignorance, i.e., with $P_*(H_i) = 0$ and $P^*(H_i) = 1$. The (41) will get us nowhere because the right-hand side is equal to zero. In the case of (42) we are reduced to the likelihood principle, i.e.,

$$(43) \qquad\qquad P^*(H_i \mid E) \approx P^*(E \mid H_i),$$

and we have made no use of indeterminacy or the apparatus of upper and lower probabilities.

Fourth, we have no conceptual basis for selecting (41) or (42), which lead to different results, even if the objections already stated, which we think are overwhelming, are overcome.

We have only quickly surveyed the obvious problems. Others will occur to the reader. We hope soon to be able to propose an alternative approach that combines Bayes' theorem and the appealing use of upper and lower probabilities to express ignorance, and at the same time is not subject to the difficulties listed above.

REFERENCES

Choquet, G.: 1955, 'Theory of Capacities', *Annales de l'Institut Fourier*, **5**, 131–295.
Dempster, A. P.: 1967, 'Upper and Lower Probabilities Induced by a Multivalued Mapping', *Annals of Mathematical Statistics*, **38**, 325–340.
Dempster, A. P.: 1968, 'A Generalization of Bayesian Inference', *Journal of the Royal Statistical Society, Series B*, **30**, 205–247.
Suppes, P.: 1972, *Axiomatic Set Theory*, Dover Publications, New York.
Suppes, P.: 1974, 'The Measurement of Belief', *Journal of the Royal Statistical Society*, Series B, **36**, 160–191.

4

Conditions on upper and lower probabilities to imply probabilities

We want to begin this paper with recording our joint indebtedness to de Finetti for so enriching the theory of probability, especially its foundations.

Suppes:

I met de Finetti in May of 1960 at a symposium in Paris on decision theory. We had several lively informal conversations about the role of the axiom of choice in the foundations of probability, especially in relation to the existence of countably additive measures. Over the years I had the opportunity to meet de Finetti on various occasions. I recall a memorable walk with de Finetti and Jimmy Savage in the gardens of the Villa Frascati near Rome later in the 1960s. On this occasion we had a long discussion of determinism and quantum mechanics, and what the existence of indeterminism and quantum mechanics implied for subjective theories of probability. The last time I saw de Finetti was in Rome in March 1979. My wife and I invited de Finetti to dinner, and after dinner he and I had a wide-ranging philosophical discussion, which he conducted with his usual vigor. In fact, even though he was then in his late seventies, I finally gave up at about 12:30 a.m., and said good night, outdone by his continuing vitality and energy in philosophical conversation.

Zanotti:

I met de Finetti in 1971 at de Finetti's summer residence in the vicinity of Rome, after correspondence beginning in 1969. He stated in one of the letters and in discussion his sympathy for the main aspects of Quine's philosophy. I especially appreciate his subjective but pluralistic view towards the foundations and application of probability, as opposed to the rigidity and limitations of many objective theories of probability. All of my thinking about probability has been strongly influenced by de Finetti.

In this paper we prove four theorems about the existence of a probability measure when a pair of upper and lower probabilities satisfy certain conditions. De Finetti did not raise this problem in precisely the form we are stating it here, but in the final appendix to his two-volume treatise, *Theory of Probability* (de Finetti, 1975, Volume 2, Appendix, Section 19.3), he discusses the question of whether imprecise probabilities exist, and he gives several different examples

Reprinted from *Erkenntnis* **31** (1989), 323–345.

in which one may want to express oneself in terms of an imprecise evaluation, as expressed by an upper and lower probability rather than a probability itself. In this section and as far as we have been able to determine elsewhere in the treatise as well, he does not raise, however, the question of whether, given a pair of upper and lower probabilities, does a probability exist that is bounded by the upper and lower probabilities. His discussion in Section 19.3 of the Appendix implicitly assumes that such a probability always does exist. It is the purpose of our paper to show some conditions under which a probability does exist. In the final theorem of the four we are also concerned with under what conditions a unique probability measure can be said to generate the upper and lower probability measures. This last theorem is related to the earlier work of Dempster (1967).

The results in this paper extend and go beyond those given in our earlier paper on using random relations to generate upper and lower probabilities (Suppes and Zanotti 1977). On the other hand, an essential tool of the mathematical constructions used in the present paper we used in two earlier papers on proving under what conditions a qualitative probability relation or a conditional probability qualitative relation has a unique numerical representation by a probability measure (Suppes and Zanotti 1976, 1982). The device was to move from the Boolean algebra of events to the semigroup of extended indicator functions generated by these events. The application is different here but it is apparent, from the rather different kinds of results obtained earlier, that the semigroup of extended indicator functions is a useful structure for studying various foundational questions in probability (the formal definitions of these concepts are given below).

1. THE FIRST THEOREM

The first theorem we prove is related to a theorem proved by Dana Scott (1964), where he gives what is now his well-known result on necessary and sufficient conditions for a qualitative ordering on the Boolean algebra of a finite set to have a representation by a probability measure. To accentuate the four main theorems we label all other statements that are proved as lemmas, propositions, or corollaries. We use repeatedly the abbreviation *iff* for *if and only if*.

DEFINITION 1 *Let Ω be a set and $\mathcal{B}$ a Boolean algebra of subsets of Ω. A pair of functions $P_*: \mathcal{B} \rightarrow [0, 1]$ and $P^*: \mathcal{B} \rightarrow [0, 1]$ is an upper–lower functional on $(\Omega, \mathcal{B})$ iff the pair satisfies the following properties for all A in $\mathcal{B}$:*

(i) $P_*(\emptyset) = P^*(\emptyset) = 0$
   where $\emptyset$ is the empty set,

30

(ii) $P_*(\Omega) = P^*(\Omega) = 1$,

(iii) $P_*(A) \leqslant P^*(A)$.

We write $G(\mathcal{B})$ for the additive semigroup generated by the indicator functions of the subsets of Ω in $\mathcal{B}$. The elements of the additive semigroup are then just what we have called earlier extended indicator functions, that is, functions that are finite sums of indicator functions. We shall also follow the notation used in our earlier papers referred to above and denote the indicator function of a set A by A^c. In what follows we shall be almost entirely concerned with finite families of sets and finite partitions. This finite restriction is in the spirit of de Finetti's concentration on finite additivity. It is apparent, but we want to emphasize that the notation $\sum A_i^c$ is a notation for a (finite) sum of indicator functions, and such a sum is of course an extended indicator function.

In the first lemma we use the following compact notation:

$$(\cup A_i)^c = 1 \wedge \left(\sum A_i^c \right) = \left\{ \omega \in \Omega : \sum A_i^c(\omega) \geqslant 1 \right\}^c$$

$$(\cup(A_i \cap A_j))^c = 2 \wedge \left(\sum A_i^c \right) = \left\{ \omega \in \Omega : \sum A_i^c(\omega) \geqslant 2 \right\}^c$$

$$\cdots$$

$$(\cap A_i)^c = n \wedge \left(\sum A_i^c \right) = \left\{ \omega \in \Omega : \sum A_i^c(\omega) \geqslant n \right\}^c,$$

where the $\Sigma, \cup, \cap$ are for $i = 1, \ldots, n, i, j = 1, \ldots, n$, etc.

LEMMA 1 *For any two finite families of sets $\{A_i\}$ and $\{B_i\}$,*

$$\sum_m A_i^c = \sum_n B_i^c \qquad \text{iff for } j \geqslant 1, \ j \wedge \sum_m A_i^c = j \wedge \sum_n B_i^c.$$

Hereafter we drop the summation indices m and n, which will be understood. Note that it is convenient to talk about representations of an extended indicator function in terms of the summation of what can be different indicator functions. It is obvious that in general an extended indicator function can be represented in many different ways as a sum of indicator functions. We call a particular family a representation.

COROLLARY $A^c = \sum B_i^c$ *iff $\{B_i\}$ is a finite partition of A.*

Given an upper–lower functional (P_*, P^*) on $(\Omega, \mathcal{B})$ we may define the following pair of functionals (F_*, F^*) on the additive semigroup $G(\mathcal{B})$:

$$F_*(f) = \text{SUP}\left\{ \sum P_*(E_i) : \sum E_i^c = f \right\}$$

$$F^*(f) = \text{INF}\left\{ \sum P^*(E_i) : \sum E_i^c = f \right\}$$

for f in $G(\mathcal{B})$, where SUP and INF are taken over all possible representations of f. Elementary properties of the pair (F_*, F^*) are stated in the next lemma.

31

LEMMA 2 *The pair (F_*, F^*) satisfies the following properties for all f, g in $G(\mathcal{B})$ and E in $\mathcal{B}$:*

(1) $0 \leqslant F_*(f) < \infty$,
(2) $F_*(f + g) \geqslant F_*(f) + F_*(g)$,
(3) $P_*(E) \leqslant F_*(E)$,
(4) $0 \leqslant F^*(f) < \infty$,
(5) $F^*(f + g) \leqslant F^*(f) + F^*(g)$,
(6) $P^*(E) \geqslant F^*(E)$.

Proof. Straightforward.

LEMMA 3 *The restriction of (F_*, F^*) to $\mathcal{B}$ will be (P_*, P^*) iff for all A, B in $\mathcal{B}$ with $A \cap B = \emptyset$ we have:*

(i) $P_*(A \cup B) \geqslant P_*(A) + P_*(B)$,
(ii) $P^*(A \cup B) \leqslant P^*(A) + P^*(B)$.

Proof. Clearly, $P_*(E) \leqslant F_*(E)$. If $\sum B_i^c = E^c$, then from the corollary $\{B_i\}$ is a partition of E. By (i) we have $P_*(E) \geqslant \sum P_*(B_i)$, hence

$$P_*(E) \geqslant \mathrm{SUP}\left\{ \sum P_*(B_i) : \sum B_i^c = E^c \right\} = F_*(E^c).$$

Conversely, from $P_*(E) = F_*(E^c) = \mathrm{SUP}\{\sum P_*(B_i) : \{B_i\}$ a partition of $E\}$ we have $P_*(E) \geqslant \sum P_*(B_i)$ for each partition $\{B_i\}$ of E. Similarly for P^*.

LEMMA 4 *Given an upper–lower functional (P_*, P^*), $F_*(f) \leqslant F^*(f)$ iff for any two representations of extended indicator function $f = \sum A_i^c = \sum B_i^c$, we have*

$$\sum P_*(A_i) \leqslant \sum P^*(B_i).$$

Proof. Suppose $F_*(f) \leqslant F^*(f)$. Let $f = \sum A_i^c = \sum B_i^c$. Then we have

$$\sum P_*(A_i) \leqslant F_*(f) \leqslant F^*(f) \leqslant \sum P^*(B_i).$$

Conversely, suppose that for any two representations $\sum A_i^c = \sum B_i^c = f$, $\sum P_*(A_i) \leqslant \sum P^*(B_i)$. Then by letting SUP and INF range over all representations we have:

$$F_*(f) = \mathrm{SUP}\left\{ \sum P_*(A_i) : \sum A_i^c = f \right\}$$
$$\leqslant \mathrm{INF}\left\{ \sum P^*(B_i) : \sum B_i^c = f \right\} = F^*(f).$$

We now define the important concept of a separating probability.

32

DEFINITION 2 *A probability P separates an upper–lower functional* (P_*, P^*) *on* $(\Omega, \mathcal{B})$ *iff for any A in* $\mathcal{B}$, $P_*(A) \leqslant P(A) \leqslant P^*(A)$.

The next lemma simply restates a theorem of P. Kranz, which is formulated for arbitrary semigroups. As might be expected, the proof of this lemma depends on the Hahn–Banach theorem.

LEMMA 5 (P. Kranz, 1972) *Let on a semigroup S be defined two real functionals U and L*: $S \to [-\infty, \infty]$ *at least one of which is finite, and such that*

  (i) $U(s + r) \leqslant U(s) + U(r)$ *for s, r in S,*
  (ii) $L(s + r) \geqslant L(s) + L(r)$ *for s, r in S,*
  (iii) $L(s) \leqslant U(s), s$ *in S.*

Then there exists an additive functional ξ on S such that

$$L(s) \leqslant \xi(s) \leqslant U(s) \text{ for all } s \text{ in } S.$$

We are now in a position to state and prove Theorem 1.

THEOREM 1 *There exists a probability P separating an upper–lower functional* (P_*, P^*) *iff for any two representations* $f = \sum A_i^c$ *and* $f = \sum B_i^c$,

$$\sum P_*(A_i) \leqslant \sum P^*(B_i).$$

Proof. For the necessity, suppose P separates (P_*, P^*). Let $f = \sum A_i^c$ and $f = \sum B_i^c$. Then (by the uniqueness of expectation of indicator functions)

$$\sum P_*(A_i) \leqslant \sum P(A_i) = \sum P(B_i) \leqslant \sum P^*(B_i).$$

The sufficiency follows from Lemma 5, a version of the Hahn–Banach theorem for abelian semigroups. From Lemma 4 we have $F_*(f) \leqslant F^*(f)$ and from Lemma 2, F_* is superadditive and F^* is subadditive. By Lemma 5 there is an additive functional F on $G(\mathcal{B})$ such that

$$F_*(f) \leqslant F(f) \leqslant F^*(f) \qquad \text{for all } f \text{ in } G(\mathcal{B}).$$

It remains to observe that F restricted to $\mathcal{B}$ is a probability. Since $F_*(\Omega^c) = F^*(\Omega^c) = 1$ and $F_*(\emptyset^c) = F^*(\emptyset^c) = 0$, we have $F(\Omega^c) = 1$ and $F(\emptyset^c) = 0$. If A and B are disjoint sets, then since $(A \cup B)^c = A^c + B^c$ we have $F((A \cup B)^c) = F(A^c) + F(B^c)$.

The form of Kranz's theorem (Lemma 5) suggests that a sufficient condition for a separating probability would be the condition: if $A \cap B = \emptyset$, then

$$P_*(A) + P_*(B) \leqslant P_*(A \cup B) \leqslant P^*(A \cup B) \leqslant P^*(A) + P^*(B),$$

but even with the additional constraint that $P^*(A) = 1 - P_*(\bar{A})$, a counterexample consisting of a Boolean algebra with seven atoms has been given by Walley (1981) and generalized by Papamarcou and Fine (1986).

33

2. THE SECOND THEOREM

The second theorem is a characterization in terms of the INF and SUP of a nonempty family of probability measures. Notice that in this case we in general get an entire family of separating probability measures.

DEFINITION 3 (P_*, P^*) *is an (upper–lower) envelope of probabilities iff there is a nonempty family* $\mathcal{P}$ *of probability measures on* $(\Omega, \mathcal{B})$ *such that for each A in* $\mathcal{B}$

$$P_*(A) = \text{INF}\{P(A): P \in \mathcal{P}\}$$

and

$$P^*(A) = \text{SUP}\{P(A): P \in \mathcal{P}\}.$$

THEOREM 2 (P_*, P^*) *is an envelope of probabilities iff for any pair of representations* $f = \sum n_i A_i^c$ *and* $f = \sum m_i B_i^c$ *we have*

(i) $\sum n_i P_*(A_i) \leqslant m_{\pi(j)} P_*(B_{\pi(j)}) + \sum_{i \neq j} m_{\pi(i)} P^*(B_{\pi(i)}),$

(ii) $\sum n_i P^*(A_i) \geqslant m_{\pi(j)} P^*(B_{\pi(j)}) + \sum_{i \neq j} m_{\pi(i)} P_*(B_{\pi(i)}),$

where π *is any permutation of the indices.*

Proof. Suppose (P_*, P^*) satisfies (i) and (ii). For C in $\mathcal{B}$ define $Q_*(A) = P_*(A)$ if A is in $\mathcal{B} - \{C\}$ else $P^*(C)$. The pair (Q_*, P^*) is an upper–lower functional with $P_* \leqslant Q_* \leqslant P^*$. Furthermore, for $f = \sum A_i^c = nC^c + \sum B_i^c$, with $B_i \neq C$, we have by (ii):

$$nP^*(C) + \sum P_*(B_i) \leqslant \sum P^*(A_i),$$

that is

$$nQ_*(C) + \sum Q_*(B_i) \leqslant \sum P^*(A_i).$$

Then by Theorem 1 there exists a probability measure P on $(\Omega, \mathcal{B})$ with:

$$P_*(A) \leqslant Q_*(A) \leqslant P(A) \leqslant P^*(A) \quad \text{for all } A \text{ in } \mathcal{B}$$

and $P(C) = P^*(C)$. Similarly define $U^*(A) = P^*(A)$ if A is in $\mathcal{B} - \{C\}$ else $P_*(C)$. The pair (P_*, U^*) is an upper–lower functional with $P_* \leqslant U^* \leqslant P^*$. A similar argument using (i) and Theorem 1 produces a probability P on $(\Omega, \mathcal{B})$ with

$$P_*(A) \leqslant P(A) \leqslant P^*(A) \quad \text{for all } A \text{ in } \mathcal{B}$$

34

and $P(C) = P_*(C)$. This process as C ranges over $\mathcal{B}$ generates a family $\mathcal{P}$ of separating probability measures for (P_*, P^*) on $(\Omega, \mathcal{B})$. The family $\mathcal{P}$ enjoys the following property for all A in $\mathcal{B}$:

$$P_*(A) = \text{INF}\{P(A): P \in \mathcal{P}\},$$
$$P^*(A) = \text{SUP}\{P(A): P \in \mathcal{P}\},$$

This proves sufficiency.

To show condition (i) is necessary, suppose

$$f = \sum n_i A_i^c = \sum m_i \boldsymbol{B}_i^c.$$

For each P in $\mathcal{P}$ and index j, we have:

$$\sum n_i P(A_i) \leqslant m_j P(B_i) + \text{SUP}\left\{ \sum_{i \neq j} m_i P(B_i) : P \in \mathcal{P} \right\}.$$

Taking INF of both sides we obtain:

$$\text{INF}\left\{ \sum n_i P(A_i): P \in \mathcal{P} \right\} \leqslant m_j \text{INF}\{P(B_j) : P \in \mathcal{P}\}$$
$$+ \text{SUP}\left\{ \sum_{i \neq j} P(B_i) : P \in \mathcal{P} \right\}.$$

From the subadditivity and superadditivity of SUP and INF, respectively, we have:

$$\sum n_i \text{INF}\{P(A_i): P \in \mathcal{P}\} \leqslant m_j \text{INF}\{P(B_j): P \in \mathcal{P}\}$$
$$+ \sum_{i \neq j} m_i \text{SUP}\{P(B_i): P \in \mathcal{P}\}$$

or

$$\sum n_i P_*(A_i) \leqslant m_j P_*(B_j) + \sum_{i \neq j} m_i P^*(B_i), \quad \text{for each index } j.$$

The necessity of condition (ii) follows from a similar argument.

It is easy to give an example of an upper–lower functional (P_*, P^*) that is not an envelope of probabilities. Let $A_i, i = 1, \ldots, 4$ be four pairwise disjoint events such that $P_*(A_i) = 0$, and $P_*(A_i \cup A_j) = P^*(A_i \cup A_j) = \frac{1}{2}, i \neq j$. Suppose now that (P_*, P^*) is an envelope. Then there must be a measure P such that $P(A_1) = 0$, $P(A_1 \cup A_2) = \frac{1}{2}$, so $P(A_2) = \frac{1}{2}$, and by similar argument $P(A_3) = P(A_4) = \frac{1}{2}$, so that $P(A_2 \cup A_3 \cup A_4) = \frac{3}{2}$, which is a contradiction.

35

3. The Third Theorem

The third theorem uses the concept of capacity of order two.

DEFINITION 4 *Let $f \in G(\mathcal{B})$. For each positive integer λ, let $E_\lambda = \{\omega \in \Omega : f(\omega) \geqslant \lambda\}$. (The set E_λ decreases when λ increases.) The representation $f = \sum E_\lambda^c$ is the spectral representation of f.*

LEMMA 6 *Let $\sum A_i^c$ be any representation of $f \in G(\mathcal{B})$. Then*

$$\left(\bigcup_i A_i \right)^c + \left(\bigcup_{i<j} (A_i \cap A_j) \right)^c + \cdots + \left(\bigcap_i A_i \right)^c$$

is the spectral representation of f.

Proof. It suffices to observe that

$$E_1 = \bigcup_i A_i, \quad E_2 = \bigcup_{i<j} (A_i \cap A_j), \text{ etc.}$$

DEFINITION 5 *The upper–lower functional (P_*, P^*) being given on $(\Omega, \mathcal{B})$, the lower integral and upper integral of $f \in G(\mathcal{B})$ are the finite positive numbers*

$$\mu_*(f) = \sum_{\lambda > 0} P_*(E_\lambda),$$

$$\mu^*(f) = \sum_{\lambda > 0} P^*(E_\lambda),$$

respectively.

LEMMA 7 *The pair (μ_*, μ^*) satisfies the following properties for all f in $G(\mathcal{B})$ and E in $\mathcal{B}$:*

(1) $0 \leqslant \mu_*(f) < \infty$,
(2) $P_*(E) = \mu_*(E)$,
(3) $0 \leqslant \mu^*(f) < \infty$,
(4) $P^*(E) = \mu^*(E)$,
(5) $\mu_*(f) \leqslant \mu^*(f)$.

DEFINITION 6 *An upper–lower functional (P_*, P^*) on $(\Omega, \mathcal{B})$ is a capacity of order two iff for all A_1 and A_2 in $\mathcal{B}$*

(i) $P_*(A_1 \cup A_2) + P_*(A_1 \cap A_2) \geqslant P_*(A_1) + P_*(A_2)$,
(ii) $P^*(A_1) = 1 - P_*(\bar{A}_1)$.

The concept of capacity has been extensively studied by Choquet (1955).

36

LEMMA 8 *If the upper–lower functional (P_*, P^*) is a capacity of order two, then for all A_1, A_2 in $\mathcal{B}$ we have:*

(i) $P_*(A_1) + P^*(A_2) \geqslant P_*(A_1 \cup A_2) + P_*(A_1 \cap A_2)$,
(ii) $P_*(A_1) + P^*(A_2) \leqslant P^*(A_1 \cup A_2) + P^*(A_1 \cap A_2)$.

Proof. Using Definition 6, the proof is straightforward.

Note that the example of an upper–lower functional (P_*, P^*) given above that is not an envelope does have properties (i) and (ii) of Lemma 8 when $A_1 \cap A_2 = \emptyset$.

LEMMA 9 *If the upper–lower functional (P_*, P^*) is a capacity of order two, then for each subfamily $\{A_i\}$ of $\mathcal{B}$ we have:*

$$(i) \quad \sum P^*(A_i) \geqslant P^*\left(\bigcup_i A_i\right) + P^*\left(\bigcup_{i \neq j}(A_i \cap A_j)\right) + \cdots + P^*\left(\bigcap_i A_i\right),$$

$$(ii) \quad \sum P_*(A_i) \leqslant P_*\left(\bigcup_i A_i\right) + P_*\left(\bigcup_{i \neq j}(A_i \cap A_j)\right) + \cdots + P_*\left(\bigcap_i A_i\right).$$

Proof. For (i) we proceed by induction on n. For the case $n = 2$ it follows from the definition of capacity of order two that:

$$P^*(A_1) + P^*(A_2) \geqslant P^*(A_1 \cup A_2) + P^*(A_1 \cap A_2).$$

The result for $n + 1$ follows from the case $k \leqslant n$ by repeated applications of the case $n = 2$ to the following inequality:

$$\sum_{i=1}^{n+1} P^*(A_i) \geqslant P^*\left(\bigcup_{i=1}^{n} A_i\right) + P^*\left(\bigcup_{i \neq j}^{n}(A_1 \cap A_j)\right)$$
$$+ \cdots + P^*\left(\bigcap_{i=1}^{n} A_i\right) + P^*(A_{n+1}).$$

Similarly for (ii).

LEMMA 10 *If the upper–lower functional (P_*, P^*) is a capacity of order two, then for each subfamily $\{A_i\}$ of $\mathcal{B}$ we have:*

$$(i) \quad P_*(A_j) + \sum_{i \neq j} P^*(A_i) \geqslant P_*\left(\bigcup_i A_i\right) + P_*\left(\bigcup_{i<j}(A_i \cap A_j)\right)$$
$$+ \cdots + P_*\left(\bigcap_i A_i\right)$$

37

$$(ii)\ \ P^*(A_j) + \sum_{i \neq j} P_*(A_i) \leqslant P^* \left(\bigcup_i A_i \right) + P^* \left(\bigcup_{i \neq j} (A_i \cap A_j) \right)$$
$$+ \cdots + P^* \left(\bigcap_i A_i \right).$$

Proof. We will prove (i); (ii) follows similarly. From Lemma 9 we can write:

$$P_*(A_{n+1}) + \sum_{i=1}^n P^*(A_i) \geqslant P^* \left(\bigcup_{i=1}^n A_i \right) + P^* \left(\bigcup_{i \neq j} (A_i \cap A_j) \right)$$
$$+ \cdots + P^* \left(\bigcap_{i=1}^n A_i \right) + P_*(A_{n+1}).$$

The result follows by repeated applications of property (i) in Lemma 8.

LEMMA 11 *For each $f \in G(\mathcal{B})$ we have*

(i) $F_*(f) = \mu_*(f),$
(ii) $F^*(f) = \mu^*(f),$

iff (P_*, P^*) is a capacity of order two.

Proof. If $\sum A_i^c$ is any representation of f, it follows from Lemma 9 that

$$\sum P_*(A_i) \leqslant P_* \left(\bigcup_i A_i \right) + P_* \left(\bigcup_{i \neq j} (A_i \cap A_j) \right)$$
$$+ \cdots + P_* \left(\bigcap_i A_i \right) = \mu_*(f).$$

Then

$$F_*(f) = \text{SUP} \left\{ \sum P_*(A_i) : \sum A_i^c = f \right\} \leqslant \mu_*(f).$$

Clearly $F_*(f) \geqslant \mu_*(f)$. Similarly, $F^*(f) = \mu^*(f)$. We have thus proved the following:

PROPOSITION 1 *The functionals μ_* and μ^* will be superadditive and subadditive respectively iff the upper–lower functional (P^*, P_*) is a capacity of order two.*

We now prove the main theorem of this section.

THEOREM 3 *If the upper–lower functional (P_*, P^*) is a capacity of order two, then it is an envelope.*

Proof. Let $\sum A_i^c$ and $\sum B_i^c$ be two representations for $f \in G(\mathcal{B})$ with $A = A_i$ for $i \leqslant m$. Then from Lemma 10 and the fact that the spectral representation of $f = mA^c$ is mA^c, we have:

$$
mP_*(A) + \sum_{i>m} P^*(A_i) \geqslant P_*\left(\bigcup_i A_i\right) + P_*\left(\bigcup_{i \neq j}(A_i \cap A_j)\right)
$$
$$
+ \cdots + P_*\left(\bigcap_i A_i\right)
$$
$$
= P_*\left(\bigcup_i B_i\right) + P_*\left(\bigcup_{i \neq j}(B_i \cap B_j)\right)
$$
$$
+ \cdots + P_*\left(\bigcap_i B_i\right)
$$
$$
\geqslant \sum_i P_*(B_i), \quad \text{by Lemma 9.}
$$

Similarly

$$
mP^*(A) + \sum_{i>m} P_*(A_i) \leqslant \sum_i P^*(B_i).
$$

DEFINITION 7 *Let $\mathcal{A}$ be a Boolean algebra on Ω, with $\mathcal{B}$ a subalgebra of $\mathcal{A}$. The inner–outer probability functional (P_*, P^*) on the probability space $(\Omega, \mathcal{B}, P)$ is the pair of set functions on $\mathcal{A}$:*

$$
P^*(A) = \mathrm{INF}\{P(B); A \subseteq B \in \mathcal{B}\},
$$
$$
P_*(A) = \mathrm{SUP}\{P(B); A \supseteq B \in \mathcal{B}\},
$$

where A is in $\mathcal{A}$.

It is clear from this definition that when A is in $\mathcal{B}$, $P_*(A) = P^*(A) = P(A)$.

LEMMA 12 *An inner–outer probability functional (P_*, P^*) is a capacity of order two.*

Proof. Clearly, $P^*(A) = 1 - P_*(\bar{A})$. We must show that

$$
P_*(A_1) + P_*(A_2) \leqslant P_*(A_1 \cup A_2) + P_*(A_1 \cap A_2).
$$

Only the infinite case requires proof. Fix $\epsilon > 0$ and choose B_1, B_2 in $\mathcal{B}$ such that $B_i \subseteq A_i$ and

$$
P_*(A_i) - \frac{\epsilon}{2} \leqslant P(B_i) \quad \text{for } i = 1, 2.
$$

39

Then

$$P_*(A_1) + P_*(A_2) - \epsilon \leqslant P(B_1) + P(B_2)$$
$$= P(B_1 \cup B_2) + P(B_1 \cap B_2)$$
$$\leqslant P_*(A_1 \cup A_2) + P_*(A_1 \cap A_2).$$

PROPOSITION 2 [Kelly] *Let $(\Omega, \mathcal{B}, P)$ be a finitely additive probability space. Let $\mathcal{A}$ be a Boolean algebra of subsets of Ω containing $\mathcal{B}$. Then P can be extended to $\mathcal{A}$.*

Proof. It follows from Lemma 12 and Theorem 3 .

This extension theorem for finitely additive probability measures shows an advantage of de Finetti's approach to probability theory, because the extension can always be made, which is not always possible for countably additive measures. This theorem is well known. We included it here as an application of Theorem 3, which leads to a very simple proof.

4. THE FOURTH THEOREM

In the first part of this section, we follow the sequence of concepts developed in Suppes and Zanotti (1977).

Let X and Y be two nonempty sets. Then the set $\mathbf{R}(X, Y)$ is the set of all (binary) relations $R \subseteq X \times Y$. We shall also occasionally refer to such a relation R as a multivalued mapping from X into Y, which is the terminology used by Dempster (1967). It is obvious that $\mathbf{R}(X, Y)$ is a Boolean algebra under the operations of intersection, union and complementation. The *domain* of a relation R is defined as

(1) $$\mathcal{D}(R) = \{x : (\exists y)(xRy)\},$$

and the notion of *range* is defined similarly.

(2) $$\mathcal{R}(R) = \{y : (\exists x)(xRy)\}.$$

The domain function $\mathcal{D}$ may also be thought of as a mapping from $\mathbf{R}(X, Y)$ to the power set, $\mathcal{P}(X)$, of X, and the range function as a mapping from $\mathbf{R}(X, Y)$ to $\mathcal{P}(Y)$.

Because of the symmetry in the domain and range mappings, we list explicitly only the properties of the domain mapping:

(3) $\mathcal{D}(\emptyset) = \emptyset$, where $\emptyset$ is the empty set, which is also the empty relation,
(4) $\mathcal{D}(U) = X$, where $U = X \times Y$ is the universal relation,
(5) $\mathcal{D}(R_1 \cup R_2) = \mathcal{D}(R_1) \cup \mathcal{D}(R_2)$, for $R_1, R_2 \in \mathbf{R}(X, Y)$,

(6) $\mathcal{D}(R_1 \cap R_2) \subseteq \mathcal{D}(R_1) \cap \mathcal{D}(R_2),$
(7) $\mathcal{D}(R_1) \sim \mathcal{D}(R_2) \subseteq \mathcal{D}(R_1 \sim R_2),$ where $\sim$ is set difference.

For several purposes it is convenient to have a restricted form of complementation: For $R \in \mathbf{R}(X, Y)$ the complement $\neg R$ is with respect to $X \times Y$, i.e.,

(8) $$\neg R = (X \times Y) \sim R,$$

the complement of $A \subseteq X$ is $X \sim A$, and the complement of $B \subseteq Y$ is $Y \sim B$. Thus $\neg \mathcal{D}(R) = X \sim \mathcal{D}(R)$. The point to note is that unrestricted complementation of sets is of no interest in the present context, i.e., it is of no interest to have the complementation of $R \in \mathbf{R}(X, Y)$ and $\mathcal{D}(R)$ relative to the same universe.

We next turn to some familiar operations on relations, or on relations and sets. The *converse* or *inverse* of a relation is defined as

(9) $$\breve{R} = \{(y, x) : xRy\}.$$

This notion is, of course, the relational generalization of function inverse. Familiar properties for R, R_1, and R_2 in $\mathbf{R}(X, Y)$ are these:

(10) $$\breve{\breve{R}} = R,$$
(11) $$\overbrace{R_1 \cap R_2} = \breve{R}_1 \cap \breve{R}_2,$$
(12) $$\overbrace{R_1 \cup R_2} = \breve{R}_1 \cup \breve{R}_2,$$
(13) $$R_1 \sim R_2 = \breve{R}_1 \sim \breve{R}_2.$$

The notion of a relation's domain being restricted to a given set is defined as

(14) $$R \mid A = R \cap (A \times \mathcal{R}(R)).$$

We next turn to two concepts that are especially important for subsequent developments. The first, $R``A$, is ordinarily called the *image* of A under the relation R, but, for reasons that will soon be made clear, we shall call it the *upper image* of A under R. The definition is simple in terms of restriction and range.

(15) $$R``A = \mathcal{R}(R|A),$$

but more suggestive is the equivalence

(16) $$y \in R``A \quad \text{iff} \quad (\exists x)(xRy \text{ and } x \in A).$$

Now let us define, for immediate comparison, the less standard notation of *lower image*, introduced in analogy to the relation between upper and lower probabilities.

(17) $$P_*(A) = 1 - P^*(\neg A).$$

41

Thus, we have

(18) $$R_{..}A = \neg(R``\neg A).$$

The 'outside' complementation of (18) is with respect to Y, and the 'inside' one with respect to X. In order to have, again in analogy to the case of upper and lower probabilities, the inequality corresponding to

(19) $$P_*(A) \leqslant P^*(A),$$

we need for the range of R to be Y, and in the case of the inverse image, the range of $\check{R}$ to be X.

(20)   If $\mathcal{R}(R) = Y$ then $R_{..}A \subseteq R``A.$

(21)   If $\mathcal{R}(\check{R}) = X$ then $\check{R}_{..}B \subseteq \check{R}``B.$

This restriction is a natural one, for it corresponds to a multivalued mapping having all of X as its domain, a point that is expanded on below.

The familiar superadditive and subadditive properties of upper and lower probabilities are expressed in the inequalities: For $A \cap B = \emptyset$,

(22) $P_*(A) + P_*(B) \leqslant P_*(A \cup B) \leqslant P^*(A \cup B) \leqslant P^*(A) + P^*(B).$

As the relational analogue we have:

(23) $(R_{..}A) \cup (R_{..}B) \subseteq R_{..}(A \cup B),$

(24) $R``(A \cup B) = (R``A) \cup (R``B),$

but (23) and (24) are not restricted to $A \cap B = \emptyset$. Some other properties of the upper and lower images of a set are the following:

(25) $R_{..}(A \cap B) = (R_{..}A) \cap (R_{..}B),$

(26) $R``(A \cap B) \subseteq (R``A) \cap (R``B),$

(27)   If $A \subseteq B$ then $R``A \subseteq R``B,$

(28)   If $A \subseteq B$ then $R_{..}A \subseteq R_{..}B,$

(29) $R_{..}\emptyset = R``\emptyset = \emptyset,$

(30) $R_{..}X = R``X = \mathcal{R}(R).$

Note that in (30) $\mathcal{R}(R)$ plays the role of the universe in the image sample space. On the basis of (25) the lower image is a homomorphism with respect to the intersection of sets, and on the basis of (24) the upper image is such a mapping with respect to the union of sets.

We now turn to relations between Boolean algebras on X and Y. Given $R \in \mathbf{R}(X, Y)$ and a Boolean algebra $\mathcal{B}$ of subsets of Y, the class

(31) $\mathcal{C}_* = \{A : A \subseteq X \;\&\; (\exists B)(B \in \mathcal{B} \;\&\; \check{R}_{..}B = A)\}$

42

is a π-system of subsets of X, i.e., it is closed under intersection, and the class

(32) $$C^* = \{A : A \subseteq X \ \& \ (\exists B)(B \in \mathcal{B} \ \& \ \check{R}``B = A)\}$$

is a family of subsets of X closed under union. The classes C_* and C^* are said to be *induced* from $\mathcal{B}$ by R. If R is a function from X to Y, then C_* and C^* are Boolean algebras and $C_* = C^*$.

It is clear that C_* and C^* each generate Boolean algebras on X, by adding closure under complementation. We have the following:

LEMMA 13 *Let $\mathcal{B}(C_*)$ and $\mathcal{B}(C^*)$ be the Boolean algebras on X generated by C_* and C^*, respectively. Then*

(33) $$\mathcal{B}(C_*) = \mathcal{B}(C^*).$$

We next introduce the concept of measurable relation, which is a natural generalization of the standard concept of measurable function. Recall first that a measurable space $(X, \mathcal{B})$ consists of a nonempty set X and a Boolean algebra of subsets of X. Given two measurable spaces $(X, \mathcal{B}_1)$ and $(Y, \mathcal{B}_2)$, a relation $R \in \mathbf{R}(X, Y)$ is said to be $(\mathcal{B}_1, \mathcal{B}_2)$-*measurable* if $\check{R}_{``}\mathcal{B}_2$ and $\check{R}``\mathcal{B}_2$ are contained in $\mathcal{B}_1$. Here extension of the upper and lower image notation to families of sets is obvious; e.g.,

(34) $$\check{R}_{``}\mathcal{B}_2 = \{A : A \subseteq X \ \& \ (\exists B)(B \in \mathcal{B}_2 \ \& \ \check{R}_{``}B = A)\}.$$

We then have the following lemma.

LEMMA 14 *In order that $R \in \mathbf{R}(X, Y)$ be $(\mathcal{B}_1, \mathcal{B}_2)$-measurable it suffices that either $\check{R}_{``}\mathcal{B}_2 \subseteq \mathcal{B}_1$ or $\check{R}``\mathcal{B}_2 \subseteq \mathcal{B}_1$.*

Given a measurable space $(Y, \mathcal{B}_2)$, a probability space $\mathcal{X} = (X, \mathcal{B}_1, P)$ and a $(\mathcal{B}_1, \mathcal{B}_2)$-measurable relation $\mathbf{R} \in \mathbf{R}(X, Y)$, we define for $A \in \mathcal{B}_2$

(35)
$$P_*(A) = P(\check{R}_{``}A)$$
$$P^*(A) = P(\check{R}_{``}A)$$

We call the pair (P_*, P^*) a *Dempsterian* functional (generated by $\mathcal{X}$ and $\mathbf{R}$) after Dempster (1967).

An upper–lower functional (P_*, P^*) on a measurable space $(Y, \mathcal{B})$ is said to be a *capacity of order n* iff the following conditions are satisfied for all $A, A_1, \ldots, A_n$ in $\mathcal{B}$:

(i) $P_*(A) - \sum_i P_*(A \cap A_i) + \sum_{i<j} P_*(A \cap A_i \cap A_j)$
$$+ \cdots + (-1)^n P_*(A \cap A_1 \cap \cdots \cap A_n) \geqslant 0.$$

(ii) $P^*(A) = 1 - P_*(\neg A)$.

43

Obviously, if (P_*, P^*) is a capacity of order n, then it is a capacity of order $m \leqslant n$. In addition, we say that (P_*, P^*) is a capacity of *infinite order* if it is a capacity of order n for all $n \geqslant 1$. As mentioned earlier, the concept of capacity is thoroughly studied by Choquet (1955). We have the following proposition relating Dempsterian functionals and capacities of infinite order.

PROPOSITION 3 *Given a measurable space* $(Y, \mathcal{B}_2)$, *a probability space* $\mathcal{X} = (X, \mathcal{B}_1, P)$, *and a* $(\mathcal{B}_1, \mathcal{B}_2)$-*measurable relation* $\mathbf{R} \in \mathbf{R}(X, Y)$ *then the Dempsterian functional* (P_*, P^*) *on* $(Y, \mathcal{B}_2)$ *generated by* $\mathcal{X}$ *and* $\mathbf{R}$ *is a capacity of infinite order.*

Proof. We will show that P_* is superadditive of arbitrary order. Given $A_1, \ldots, A_n$ in $\mathcal{B}_2$ we have $\check{\mathbf{R}}_{..} A_1, \ldots, \check{\mathbf{R}}_{..} A_n$ in $\mathcal{B}_1$. Then

$$P\left(\bigcup_i \check{\mathbf{R}}_{..} A_i\right) - \sum_i P(\check{\mathbf{R}}_{..} A_i) + \sum_{i<j} P(\check{\mathbf{R}}_{..} A_i \cap \check{\mathbf{R}}_{..} A_j)$$

$$+ \cdots + (-1)^n P\left(\bigcap_i \check{\mathbf{R}}_{..} A_i\right) = 0.$$

Recalling that $\cup \check{\mathbf{R}}_{..} A_i \subseteq \check{\mathbf{R}}_{..}(\cup A_i)$ and $\check{\mathbf{R}}_{..}(\cap A_i) = \cap \check{\mathbf{R}}_{..} A_i$ we have (i). For (ii)

$$P^*(A) = P(\check{\mathbf{R}}_{..} A) = P(\overline{\check{\mathbf{R}}_{..} \bar{A}})$$

$$= 1 - P(\check{\mathbf{R}}_{..} \bar{A})$$

$$= 1 - P_*(\bar{A}).$$

The converse of this proposition is true, but for the sake of simplicity we will prove it later, but only for the finite case.

DEFINITION 8 *Let* $(\Omega, \mathcal{B})$ *and* $(\mathcal{B}, \mathcal{P}(\mathcal{B}))$ *be given where* Ω *is a nonempty set,* $\mathcal{B}$ *a Boolean algebra of subsets of* Ω, *and* $\mathcal{P}(\mathcal{B})$ *the power set of* $\mathcal{B}$. *Define two set-valued set functions*

$$\mathcal{B} \xrightarrow[( )_*]{( )^*} \mathcal{P}(\mathcal{B}) \ as \ follows:$$

For each A in $\mathcal{B}$,

$$(A)_* = \{B : B \in \mathcal{B}, B \subseteq A\}$$

$$(A)^* = \{B : B \in \mathcal{B}, B \cap A \neq \emptyset\}$$

These mappings are the lower mapping *and* upper mapping *of* $\mathcal{B}$ *into* $\mathcal{P}(\mathcal{B})$, *respectively.*

44

LEMMA 15 *The lower mapping* $(\ )_*$ *satisfies the following properties:*

(1) It is injective,
(2) $A \subseteq B$ iff $(A)_* \subseteq (B)_*$, i.e., it is monotonic,
(3) $(\varnothing)_* = \{\varnothing\}$,
(4) $(\Omega)_* = \mathcal{B}$,
(5) $(A)_* \cup (B)_* \subseteq (A \cup B)_*$,
(6) $(A)_* \cap (B)_* = (A \cap B)_*$.

Thus $(\ )_*$ is a homomorphism for intersection.

The proofs of these results are immediate.

LEMMA 16 *The upper mapping* $(\ )^*$ *satisfies the following properties:*

(1) It is injective,
(2) $A \subseteq B$ iff $(A)^* \subseteq (B)^*$, i.e., it is monotonic,
(3) $(\varnothing)^* = \{\varnothing\}$,
(4) $(\Omega)^* = \mathcal{B} - \varnothing$,
(5) $(A)^* \cup (B)^* = (A \cup B)^*$,
(6) $(A)^* \cap (B)^* \supseteq (A \cap B)^*$.

Thus $(\ )^*$ is a homomorphism for union.

The proofs of these results are immediate.

For notational convenience we will write A_* and A^* for $(A)_*$ and $(A)^*$, respectively, when no confusion will arise.

LEMMA 17 *We have*

(i) $\overline{(A^*)} = (\bar{A})_*$ and $(\bar{A})^* = \overline{(A_*)}$,
(ii) $A^* \cap B^* = \varnothing$ iff $A = \varnothing$ or $B = \varnothing$,
(iii) $A_* \cup B_* \subseteq (A \cup B)_*$,
(iv) $A_* \cup B^* \subseteq (A \cup B)^*$,
(v) $(A \cup B)_* \subseteq A_* \cup B^*$,
(vi) $A_* \cap B_* = \varnothing$ iff $A \cap B = \varnothing$,
(vii) $A_* \cap B^* = \varnothing$ iff $A \cap B = \varnothing$.

In particular the following chain of inclusions is true for all A, B in $\mathcal{B}$:

$$A_* \cup B_* \subseteq (A \cup B)_* \subseteq A_* \cup B^* \subseteq (A \cup B)^* = A^* \cup B^*.$$

The proof of the lemma is lengthy but straightforward.

The following lemma gives a characterization of the set of images of the lower mappings.

LEMMA 18 *Let $\mathcal{C} = \{A_*: A \in \mathcal{B}\}$. Then $\mathcal{C}$ is a Dynkin's π-system of subsets of $\mathcal{B}$.*

Proof. If $A_*, B_* \in \mathcal{C}$ we have $A_* \cap B_* = (A \cap B)_* \in \mathcal{C}$ since $A \cap B \in \mathcal{B}$.

LEMMA 19 *Assume $\mathcal{B}$ finite and let $\mathcal{D} = \{A_*, A^*: A \in \mathcal{B}\}$. Then the Boolean algebra generated by $\mathcal{D}$ is $\mathcal{P}(\mathcal{B})$.*

Proof. For each A in $\mathcal{B}$ there is a family $\{A_i\}$ of atoms of $\mathcal{B}$ such that we can uniquely write

$$A = \bigcup_{i=1}^{n} A_i.$$

Furthermore we can write:

$$\{A\} = \left(\bigcap_{i=1}^{n} A_i^*\right) \cap \left(\bigcup_{i=1}^{n} A_i\right)_*.$$

In fact

$$\bigcup_{i=1}^{n} A_i \in A_i^*$$

for each $i \leqslant n$ and

$$\bigcup_{i \neq j} A_i \notin A_j^*$$

for each $j \leqslant n$.

DEFINITION 9 *The triple $(\Omega, \mathcal{B}, (P_*, P^*))$ consisting of a set Ω, a Boolean algebra $\mathcal{B}$ of subsets of Ω and a Dempsterian functional (P_*, P^*) on $\mathcal{B}$, is a Dempsterian space.*

We will say that $(\Omega, \mathcal{B}, (P_*, P^*))$ is a *finite* Dempsterian space if Ω is a finite set.

LEMMA 20 *The triple $(\Omega, \mathcal{B}, (P_*, P^*))$ is a finite Dempsterian space iff there is a function $f: \mathcal{B} \to [0, 1]$ satisfying the following properties:*

(i) $0 \leqslant f(A) \leqslant 1$ *for every $A \in \mathcal{B}$, and $f(\emptyset) = 0$,*

(ii) $\displaystyle\sum_{A \in \mathcal{B}} f(A) = 1.$

Proof. Given $A \in \mathcal{B}$ we have $A = \cup A_i$ with A_i's atoms of $\mathcal{B}$. Let $B_i = \cup_{j \neq i} A_j$. From superadditivity we can define f explicitly:

$$f(A) = P_*(A) - \sum_i P_*(B_i) + \sum_{i<j} P_*(B_i \cap B_j)$$

$$+ \cdots + (-1)^n P_*\left(\bigcap_i B_i\right) \geqslant 0.$$

It is simple to show that

$$\sum_{A \in \mathcal{B}} f(A) = P_*(\Omega) = 1.$$

Conversely, given a function $f: \mathcal{B} \to [0, 1]$ we can define for all A in $\mathcal{B}$

$$P_*(A) = \sum_{\substack{B \subset A \\ B \in \mathcal{B}}} f(B)$$

and

$$P^*(A) = \sum_{\substack{B \cap A \neq \emptyset \\ B \in \mathcal{B}}} f(B)$$

and show that (P_*, P^*) is a Dempsterian functional on $\mathcal{B}$. We will not do this here, since the proof of the converse follows from the following lemma:

LEMMA 21 *Given a finite measurable space $(\Omega, \mathcal{B})$ and a probability space $(\mathcal{B}, \mathcal{P}(\mathcal{B}), P)$, define for each A in $\mathcal{B}$*

$$P_*(A) = P(A_*),$$
$$P^*(A) = P(A^*).$$

The pair (P_*, P^*) is a Dempsterian functional on $\mathcal{B}$.

Proof.

$$\begin{aligned}
P^*(A) = P(A^*) &= P(\overline{(\bar{A})_*}) && \text{by Lemma 17,} \\
&= 1 - P((\bar{A})_*) \\
&= 1 - P_*(\bar{A}).
\end{aligned}$$

To $A_1, A_2, \ldots, A_n$ in $\mathcal{B}$ there correspond their images $(A_1)_*, (A_2)_*, \ldots, (A_n)_*$ in $\mathcal{P}(\mathcal{B})$. We have:

$$P\left(\bigcup_i (A_i)_*\right) - \sum_i P(A_i)_* + \sum_{i<j} P((A_i)_* \cap (A_j)_*)$$
$$+ \cdots + (-1)^n P_*\left(\bigcap_i (A_i)_*\right) = 0.$$

By Lemma 15(6) we have:

$$P\left(\bigcup_i (A_i)_*\right) - \sum_i P(A_i)_* + \sum_{i<j} P(A_i \cap A_j)_*$$
$$+ \cdots + (-1)^n P\left(\bigcap_i A_i\right)_* = 0.$$

47

From $\cup(A_i)_* \subseteq (\cup A_i)_*$, which is (iii) of Lemma 17, we have:

$$P_*(\cup A_i) - \sum_i P_*(A_i) + \sum_{i<j} P_*(A_i \cap A_j)$$

$$+ \cdots + (-1)^n P_*\left(\bigcap_i A_i\right) \geqslant 0.$$

Observe that any function $f: \mathcal{B} \to [0, 1]$ with $f(\emptyset) = 0$ and $\sum_{A \in \mathcal{B}} f(A) = 1$ defines a probability measure on $\mathcal{P}(\mathcal{B})$. Thus we have proved the following representation – the first part ((i)) of Theorem 4.

THEOREM 4 (i) *There is a* $1:1$ *correspondence between finite Dempsterian measures on* $(\Omega, \mathcal{B})$ *and probability measures on* $(\mathcal{B}, \mathcal{P}(\mathcal{B}))$.

(ii) *Given a finite Dempsterian space* $(\Omega, \mathcal{B}, (P_*, P^*))$ *there exists a probability space* $(\Omega_1, \mathcal{B}_1, P)$ *and a unique* $(\mathcal{B}_1, \mathcal{B})$-*measurable random relation* $R(\Omega_1, \Omega)$ *generating it.*

Proof. For the proof of (ii), let

$$\Omega_1 = \mathcal{B} \qquad \text{and} \qquad \mathcal{B}_1 = \mathcal{P}(\mathcal{B})$$

following our earlier construction. Then, by (i) there is a unique probability measure P on $\mathcal{B}_1$ with the property $P(A_*) = P_*(A)$ and $P(A^*) = P^*(A)$ for all A in $\mathcal{B}$. Second, we define the random relation R as follows:

$$R = \{(A, \omega) : A \in \mathcal{B} \; \& \; A \neq \emptyset \; \& \; \omega \in A\}.$$

Clearly

$$\check{R}``A = A^*$$

and

$$\check{R}_{``}A = A_*.$$

Moreover,

$$P^*(A) = P(\check{R}``A),$$

and similarly for P_*, which completes the proof.

In one clear sense, Theorem 4 is a generalization of Kolmogorov's representation theorem for random variables, i.e., if a random quantity in the sense of de Finetti has a distribution, then there exists a probability space and a real-valued function generating the distribution. We call the function a random variable, which is a representation of the random quantity. What we have done in Theorem 4(ii) is to generalize from random variables to random relations, and such

48

relations express the indeterminacy characteristic of Dempsterian functionals. Of course, this generalization has a price, e.g., the theory of upper and lower conditional probability is not entirely satisfactory (see the last section of Suppes and Zanotti 1977).

An axiomatic *qualitative* theory of random quantities is developed in Suppes and Zanotti (1992).

REFERENCES

Choquet, G.: 1955, 'Theory of Capacities', *Annales de l'Institut Fourier*, **5**, 131–295.

de Finetti, B.: 1975, *Theory of Probability*, vols. 1 and 2, Wiley, New York (English translation of 1970 Italian edition).

Dempster, A. P.: 1967, 'Upper and Lower Probabilities Induced by a Multivalued Mapping', *Annals of Mathematical Statistics*, **38**, 325–40.

Kranz, P.: 1972, 'Additive functionals on Abelian Semigroups', *Annales Societatis Mathematicae Polonae, Series I: Commentationes Mathematicae*, **XVI**, 239–46. *Rocznieki Polskiego Towarzystwa Matematycznego, Seria I: Prace Matematyczne*, **XVI**.

Papamarcou. A., and Fine, T. L.: 1986, 'A Note on Undominated Lower Probabilities', *The Annals of Probability*, **14**, 710–23.

Scott, D.: 1964, 'Measurement Models and Linear Inequalities, *Journal of Mathematical Psychology*, **1**, 233–47.

Suppes, P., and Zanotti, M.: 1976,'Necessary and Sufficient Conditions for Existence of a Unique Measure Strictly Agreeing with a Qualitative Probability Ordering', *Journal of Philosophical Logic*, **5**, 431–38, [chap. 1, this volume].

Suppes, P., and Zanotti, M.: 1977, 'On Using Random Relations to Generate Upper and Lower Probabilities', *Synthese*, **36**, 427–40 [chapter 3, this volume].

Suppes, P., and Zanotti, M.: 1982,'Necessary and Sufficient Qualitative Axioms for Conditional Probability', *Z. Wahrscheinlichkeitstheorie verw. Gebiete*, **60**, 163–69 [chap. 2, this volume].

Suppes, P., and Zanotti, M.: 1992, 'Qualitative Axioms for Random-Variable Representation of Extensive Quantities,' in C. W. Savage and P. Ehrlich (eds.), *The Nature and Purpose of Measurement*. Minneapolis: University of Minnesota Press [chap. 5, this volume].

Walley, P.: 1981, 'Coherent Lower (and Upper) Probabilities', *Technical Report*. Department of Statistics, University of Warwick, England.

5

Qualitative axioms for random-variable representation of extensive quantities

In the standard theory of fundamental extensive measurement, qualitative axioms are formulated that lead to a numerical assignment unique up to a positive similarity transformation. The central idea of the theory of *random* quantities is to replace the numerical assignment by a random-variable assignment. This means that each object is assigned a random variable. In the case of extensive quantities, the expectation of the random variable replaces the usual numerical assignment, and the distribution of the random variable reflects the variability of the property in question, which could be intrinsic to the object or due to errors of observation. In any case, the existence of distributions with positive variances is almost universal in the actual practice of measurement in most domains of science.

It is a widespread complaint about the foundations of measurement that too little has been written that combines the qualitative structural analysis of measurement procedures and the analysis of variability in a quantity measured or in errors in the procedures used. In view of the extraordinarily large number of papers that have been written about the foundations of the theory of error, which go back to the eighteenth century with fundamental work already by Simpson, Lagrange, and Laplace, followed by the important contributions of Gauss, it is surprising that the two kinds of analysis have not received a more intensive consideration. Part of the reason is the fact that, in all of this long history, the literature on the theory of errors has been intrinsically quantitative in character. Specific distributional results have usually been the objective of the analysis, and the assumptions leading to such results have been formulated in quantitative probabilistic terms. This quantitative framework is also assumed in the important series of papers by Falmagne and his collaborators on random-variable representations for interval, conjoint, and extensive measurement (see

Reprinted from C. W. Savage and P. E. Ehrlich (Eds.), *Philosophical and Foundational Issues in Measurement Theory*. Hillsdale, N. J.: Erlbaum, 1992, pp. 39–52.

Falmagne, 1976, 1978, 1979, 1980, 1985; Falmagne & Iverson, 1979; Iverson & Falmagne, 1985).

In light of this long history, we would certainly not want to claim that the various results presented in this chapter solve all the natural kinds of questions that have been in the air for some time. We do believe that we have taken a significant step toward combining in one analysis the qualitative structures characteristic of the foundations of measurement and the probabilistic structures characteristic of the theory of error or the theory of variability.

The approach to the distribution of the representing random variables of an object consists of developing, in the usual style of the theory of measurement, qualitative axioms concerning the *moments* of the distribution, which are represented as expectations of powers of the representing numerical random variable. The first natural question is whether or not there can be a well-defined qualitative procedure for measuring the moments. This is discussed in the first section. Section 2 presents the qualitative primitive concepts and Section 3 the axiom system. The representation theorem and its proof are given in Section 4.

1. VARIABILITY AS MEASURED BY MOMENTS

The approach to the distribution of the representing random variable of an object consists of developing, in the usual style of the theory of measurement, qualitative axioms concerning the *moments* of the distribution, which are represented as expectations of powers of the representing numerical random variable. The classic problem of moments in the theory of probability enters in an essential way in the developments to follow. We lay out in an explicit way the qualitative assumptions about moments that are made.

Before giving the formal developments, we address the measurement of moments from a qualitative standpoint. We outline here one approach without any claim that it is the only way to conceive of the problem. In fact, we believe that the pluralism of approaches to measuring probability is matched by that for measuring moments, for reasons that are obvious given the close connection between the two.

The one approach we outline here corresponds to the limiting relative-frequency characterization of probability, which we formulate here somewhat informally. Let s be an infinite sequence of independent trials with the outcome on each trial being heads or tails. Let $H(i)$ be the number of heads on the first i trials of s. Then, relative to s,

$$P(\text{heads}) = \lim_{i \to \infty} H(i)/i,$$

with the provision that the limit exists and that the sequence s satisfies certain conditions of randomness that need not be analyzed here. In practice, of course,

51

only a finite initial segment of any such sequence is realized as a statistical sample. However, ordinarily in the case of probability, the empirical procedure encompasses several steps. In the approach given here, the first step is to use the limiting relative-frequency characterization. The second step is to produce and analyze a finite sample with appropriate statistical methods.

Our approach to empirical measurement of qualitative moments covers the first step but not the second of giving detailed statistical methods. Thus, let a_0 be an object of small mass of which we have many accurate replicas – so we are assuming here that the variability in a_0 and its replicas, $a_0^{(j)}$, $j = 1, 2, \ldots$ are negligible. Then we use replicas of a_0 to qualitatively weigh an object a. On each trial, we force an equivalence, as is customary in classical physics. Thus, on each trial i, we have

$$a \sim \left\{ a_0^{(1)}, a_0^{(2)}, \ldots, a_0^{(m_i)} \right\}.$$

The set shown on the right we symbolize as $m_i a_0$. Then, as in the case of probability, we characterize a^n, the nth qualitative raw moment of a, by

$$a^n \sim \lim_{j \to \infty} \frac{1}{j} \sum_{i=1}^{j} m_i^n a_0,$$

but, in practice, we use a finite number of trials and use the estimate $\hat{a}^n$,

$$\hat{a}^n \sim \frac{1}{j} \sum_{i=1}^{j} m_i^n a_0,$$

and so also only estimate a finite number of moments. It is not to the point here to spell out the statistical procedures for estimating a^n. Our objective is only to outline how one can approach empirical determination of the qualitative raw moments.

There is one important observation to deal with. The observed data, summarized in the integers $m_1, m_2, \ldots, m_j$, on which the computation of the moments is based, also constitute a histogram of the distribution. Why not estimate the distribution directly? When a distribution of a particular form is postulated, there need be no conflict in the two methods, and the histogram can be of further use in testing goodness of fit.

The reason for working with the raw moments is theoretical rather than empirical or statistical. Various distributions can be qualitatively characterized in terms of their raw moments in a relatively simple way, as the examples in the Corollary to the Representation Theorem show. Furthermore, general qualitative conditions on the moments are given in the Representation Theorem. Alternative qualitative approaches to characterizing distributions undoubtedly exist and as they are developed may well supersede the one used here.

52

We now turn to the formal developments. In proving the representation theorem for random extensive quantities in this section, we apply a well-known theorem of Hausdorff (1923) on the one-dimensional moment problem for a finite interval.

HAUSDORFF'S THEOREM *Let* $\mu_0, \mu_1, \mu_2, \ldots$ *be a sequence of real numbers. Then a necessary and sufficient condition that there exist a unique probability distribution* F *on* $[0, 1]$ *such that* μ_n *is the nth raw moment of the distribution* F, *that is to say,*

$$(1) \qquad \mu_n = \int_0^1 t^n dF, \qquad n = 0, 1, 2, \ldots,$$

is that $\mu_0 = 1$ *and all the following inequalities hold:*

$$(2) \quad \mu_\nu - \binom{k}{1}\mu_{\nu+1} + \binom{k}{2}\mu_{\nu+2} + \cdots + (-1)^k \mu_{\nu+k} \geqslant 0 \text{ for } k, \nu = 0, 1, 2, \ldots$$

A standard terminology is that a sequence of numbers $\mu_n, n = 0, 1, 2, \ldots$ is *completely monotonic* iff Inequalities (2) are satisfied, in more compact binomial notation $\mu^\nu (1 - \mu)^k \geqslant 0$, for $k, \nu = 0, 2, \ldots$ (for detailed analysis of many related results on the problem of moments, see Shohat & Tamarkin, 1943).

It is important to note that we do not need an additional separate specification of the domain of definition of the probability distribution in Hausdorff's theorem. The necessary and sufficient conditions expressed in the Inequalities (2) guarantee that all the moments lie in the interval $[0, 1]$, and so this may be taken to be the domain of the probability distribution without further assumption.

2. QUALITATIVE PRIMITIVES FOR MOMENTS

The idea, then, is to provide a qualitative axiomatization of the moments for which a qualitative analogue of Inequalities (2) obtains and then to show that the qualitative moments have a numerical representation that permits one to invoke Hausdorff's theorem. Thus, the qualitative structure begins first with a set G of objects. These are the physical objects or entities to which we expect ultimately to associate random variables. More precisely, we expect to represent the selected extensive attribute of each object by a random variable. However, in order to get at the random variables, we must generate from G a set of entities that we can think of as corresponding to the raw moments and mixed moments of the objects in G. To do that, we must suppose that there is an operation $\bullet$ of combining so that we can generate elements $a^n = a^{n-1} \bullet a$, which, from a qualitative point of view, will be thought of as corresponding to the raw moments of a. It is appropriate to think of this operation as an operation of multiplication, but it corresponds to multiplication of random variables, not

53

to multiplication of real numbers. We shall assume as axioms that the opertion is associative and commutative, but that it should not be assumed to be distributive with respect to disjoint union (which corresponds to numerical addition) can be seen from the following random-variable counterexample, given in Gruzewska (1954). Let X_1, X_2, X_3 be three random variables, where

$$X_1 = X_2 = \begin{cases} 0 \\ 1 \end{cases} \text{with } P(X_2 = 0) = P(X_1 = 1) = \frac{1}{2},$$
$$X_3 = 1 \text{ with } P(X_3 = 1) = 1.$$

Then

$$X_2 + X_3 = \begin{cases} 1 \text{ with } P(X_2 + X_3 = 1) = \frac{1}{2} \\ 2 \text{ with } P(X_2 + X_3 = 2) = \frac{1}{2} \end{cases}$$

$$L_1 = X_1(X_2 + X_3) = \begin{cases} 0 \text{ with } P(L_1 = 0) = \frac{1}{2} \\ 1 \text{ with } P(L_1 = 1) = \frac{1}{2} \cdot \frac{1}{2} = \frac{1}{4} \\ 2 \text{ with } P(L_1 = 2) = \frac{1}{2} \cdot \frac{1}{2} = \frac{1}{4} \end{cases}$$

and

$$X_1 X_2 = \begin{cases} 0 \text{ with } P(X_1 X_2 = 0) = \frac{3}{4} \\ 1 \text{ with } P(X_1 X_2 = 1) = \frac{1}{4} \end{cases}$$

$$X_1 X_3 = \begin{cases} 0 \text{ with } P(X_1 X_3 = 0) = \frac{1}{2} \\ 1 \text{ with } P(X_1 X_3 = 1) = \frac{1}{2} \end{cases}$$

and

$$L_2 = X_1 X_2 + X_1 X_3 = \begin{cases} 0 \text{ with } P(L_2 = 0) = \frac{3}{8} \\ 1 \text{ with } P(L_2 = 1) = \frac{3}{8} + \frac{1}{8} = \frac{4}{8} = \frac{1}{2} \\ 2 \text{ with } P(L_2 = 2) = \frac{1}{8}. \end{cases}$$

(The computations make clear the assumptions of independence made, but the random-variable notation in this counterexample must be interpreted in terms of types, so that representative independent tokens may be selected as needed. So, for example, $X_1 = X_2$ means these two classes of random variables, i.e., types, are the same.) As can be seen, L_1 and L_2 have different distributions, although

$$E(X_1(X_2 + X_3)) = E(X_1 X_2 + X_1 X_3).$$

We turn now to the explicit definition of a semigroup that contains the associative and commutative axioms of multiplication.

DEFINITION 1 *Let A be a nonempty set, G a nonempty set, · a binary operation on A, and 1 an element of G. Then* $\mathcal{U} = (A, G, \cdot, 1)$ *is a commutative semigroup with identity 1 generated by G iff the following axioms are satisfied for every a, b, and c in G.*

1. *If $a \in G$, then $a \in A$.*
2. *If $s, t \in A$, then $(s \cdot t) \in A$.*
3. *Any member of A can be generated by a finite number of applications of Axioms 1–3 from elements of G.*
4. $a \cdot (b \cdot c) = (a \cdot b) \cdot c$.
5. $a \cdot b = b \cdot a$.
6. $1 \cdot a = a$.

Note that, because of the associativity axiom, we omit parentheses from here on. Note, further, that, on the basis of Axiom 3, we think of elements of A as finite strings of elements of G. Intuitively the elements of A are qualitative mixed moments. Furthermore, because the product operation · is associative and commutative, we can always write the mixed moments in a standard form involving powers of the generators. For example, $a \cdot a \cdot a \cdot c \cdot a \cdot b \cdot c = a^4 \cdot b \cdot c^2$. This expression is interpreted as the qualitative mixed moment consisting of the fourth raw moment of a times the first one of b times the second one of c. We denote this semigroup by A.

Our last primitive is a qualitative ordering of moments. As usual, we will denote it by $\succcurlyeq$. The first question concerns the domain of this relation. For purposes of extensive measurement, it is useful to assume that the domain is all finite subsets from the elements of the semigroup A. We may state this as a formal definition:

DEFINITION 2 *Let A be a nonempty set and $\succcurlyeq$ a binary relation on $\mathcal{F}$, the family of all finite subsets of A. Then* $\mathfrak{A} = (A, \mathcal{F}, \succcurlyeq)$ *is a weak extensive structure iff the following axioms are satisfied for every B, C, and D in $\mathcal{F}$:*

1. *The relation $\succcurlyeq$ is a weak ordering of $\mathcal{F}$.*
2. *If $B \cap D = C \cap D = \emptyset$, then $B \succcurlyeq C$ iff $B \cup D \succcurlyeq C \cup D$.*
3. *If $B \neq \emptyset$, then $B \succ \emptyset$.*

Superficially the structure just defined looks like a familiar structure of qualitative probability, but in fact it is not. The reason is that because A is an infinite set, we cannot assume $\mathcal{F}$ is closed under complementation, because that would violate the assumption that the subsets in $\mathcal{F}$ are finite.

An important conceptual point is that we do require the ordering in magnitude of different raw moments. One standard empirical interpretation of what it means to say that the second raw moment, a^2, is less than the first, a^1, was outlined previously. A formal point, appropriate to make at this stage, is to contrast the uniqueness result we anticipate for the representation theorem with the usual uniqueness up to a similarity (i.e., multiplication by a positive constant) for extensive measurement. We have, in the present setup, not only the extensive operation but also the semigroup multiplication for forming moments; therefore, the uniqueness result is absolute (i.e., uniqueness in the sense of the identity function). Given this strict uniqueness, the magnitude comparison of a^m and a^n for any natural numbers m and n is not a theoretical problem. It is of course apparent that any procedure for measurement of moments, fundamental or derived, will need to satisfy such strict uniqueness requirements in order to apply Hausdorff's or other related theorems in the theory of moments.

Within $\mathcal{F}$, we may define what it means to have n disjoint copies of $B \in \mathcal{F}$:

$$1B = B$$
$$(n + 1)B \sim nB \cup B',$$

where $nB \cap B' = \varnothing$, and $B' \sim B$ and $\sim$ is the equivalence relation defined in terms of the basic ordering $\succcurlyeq$ on $\mathcal{F}$. Axiom 3 will simply be the assumption that such a B' always exists, and so nB is defined for each n. It is essential to note that this standard extensive or additive recursive definition is quite distinct from the one for moments a^n given earlier.

3. AXIOM SYSTEM FOR QUALITATIVE MOMENTS

Our goal is to provide axioms on the qualitative raw moments such that we can prove that object a can be represented by a random variable X_a, and the nth raw moment a^n is represented by the nth power of X_a (i.e., by X_a^n).

For convenience, we shall assume the structures we are dealing with are bounded in two senses. First, the set G of objects will have a largest element 1, which intuitively means that the expectation of the random variables associated with the elements of a will not exceed that of 1. Moreover, we will normalize things so that the expectation associated with X_1 is 1. This normalization shows up in the axiomatization as 1 acting as the identity element of the semigroup. Second, because of the condition arising from the Hausdorff theorem, this choice means that all of the raw moments are decreasing in powers of n (i.e., if $m \leqslant n$, then $a^n \preccurlyeq a^m$). Obviously the theory can be developed so that the masses are greater than 1, and the moments become larger with increasing n. This is the natural theory when the probability distribution is defined on the positive real line. As might be expected, the conditions are simpler for the existence of a probability distribution on a finite interval, and this is also realistic from a methodological standpoint. The exponential notation for qualitative moments a^n is intuitively clear, but it is desirable to have the following formal recursive definition:

$$a^0 = 1,$$
$$a^n = a^{n-1} \cdot a,$$

in order to have a clear interpretation of a^0.

Before giving the axiom system, we must discuss more fully the issue of what will constitute a qualitative analogue of Hausdorff's condition, Inequality (2).

We have only an operation corresponding to addition and not to subtraction in the qualitative system; thus, for k, an even number, we rewrite this inequality solely in terms of addition as follows:

$$(3) \qquad \mu_v + \binom{k}{2}\mu_{v+2} + \cdots + \mu_{v+k} \geq \binom{k}{1}\mu_{v+1} + \cdots + k\mu_{v+(k-1)},$$

and a corresponding inequality for the case in which k is odd. In the qualitative system, the analogue to Inequality (3) must be written in terms of union of sets as follows for k even:

$$(4) \qquad a^v \cup \binom{k}{2}a^{v+2} \cup \cdots \cup a^{v+k} \succcurlyeq \binom{k}{1}a^{v+1} \cup \cdots \cup ka^{v+(k-1)}.$$

When k is odd,

$$(5) \qquad a^v \cup \binom{k}{2}a^{v+2} \cup \cdots \cup ka^{v+(k-1)} \succcurlyeq \binom{k}{1}a^{v+1} \cup \cdots \cup a^{v+k}.$$

There are several remarks to be made about this pair of inequalities. First of all, we can infer that, for $a \prec 1$, as opposed to $a \sim 1$, the moments are a strictly decreasing sequence (i.e., $a^v \succ a^{v+1}$). Second, the meaning of such terms as $\binom{k}{2}a^{v+2}$ was recursively defined earlier, with the recursion justified by Axiom 3 below. It is then easy to see that the unions indicated in Inequalities (4) and (5) are of disjoint sets. On the basis of the earlier terminology, we can then introduce the following definition. A qualitative sequence $a^0, a^1, a^2, a^3, \ldots$ is *qualitatively completely monotonic* iff Inequalities (4) and (5) are satisfied.

DEFINITION 3 *A structure* $\mathfrak{A} = (A, \mathcal{F}, G, \succcurlyeq, \cdot, 1)$ *is a random extensive structure with independent objects – the elements of G – iff the following axioms are satisfied for a in G, s and t in A, k, m, m', n, and n' natural numbers and B and C in $\mathcal{F}$:*

1. *The structure* $(A, \mathcal{F}, \succcurlyeq)$ *is a weak extensive structure.*
2. *The structure* $(A, G, \cdot, 1)$ *is a commutative semigroup with identity 1 generated by G.*

57

3. *There is a C' in $\mathcal{F}$ such that $C' \sim C$, and $C' \cap B = \varnothing$;*
4. *Archimedean. If $B \succ C$, then, for any D in $\mathcal{F}$, there is an n such that*

$$nB \succeq nC \cup D.$$

5. *Independence. Let mixed moments s and t have no common objects:*
   a. *If $m1 \succeq ns$, and $m'1 \succeq n't$, then $mm'1 \succeq nn'(s \cdot t)$*
   b. *If $m1 \preccurlyeq ns$, and $m'1 \preccurlyeq n't$, then $mm'1 \preccurlyeq nn'(s \cdot t)$*
6. *The sequence $a^0, a^1, a^2, \ldots$ of qualitative raw moments is qualitatively completely monotonic.*

The content of Axiom 1 is familiar. What is new here is, first of all, Axiom 2, in which the commutative semigroup, as mentioned earlier, is used to represent the mixed moments of a collection of objects. Axiom 3 is needed in order to make the recursive definition of $(n + 1)B$ well defined as given earlier. The special form of the Archimedean axiom is the one needed when there is no solvability axiom, as discussed in Section 3.2.1 of Krantz, Luce, Suppes, and Tversky (1971). The dual form of Axiom 5 is just what is needed to prove the independence of the moments of different objects, which means that the mixed moments factor in terms of expectation. Note that it is symmetric in $\succeq$ and $\preccurlyeq$. The notation used in Axiom 5 involves both disjoint unions, as in $m1$, and the product notation for mixed moments, as in $(s \cdot t)$. Axiom 6 formulates the qualitative analogue of Hausdorff's necessary and sufficient condition as discussed above.

4. REPRESENTATION THEOREM AND PROOF

REPRESENTATION THEOREM *Let $\mathfrak{A} = (A, \mathcal{F}, G, \succeq, \cdot, 1)$ be a random extensive structure with independent objects. Then there exists a family $\{X_B, B \in \mathcal{F}\}$ of real-valued random variables such that:*

   (i) *every object a in G is represented by a random variable $\mathbf{X}_a$ whose distribution is on $[0, 1]$ and is uniquely determined by its moments,*
   (ii) *the random variables $\{X_a, a \in G\}$ are independent,*
   (iii) *for a and b in G, with probability one,*

$$\mathbf{X}_{a \cdot b} = \mathbf{X}_a \cdot \mathbf{X}_b,$$

   (iv) *$E(X_B) \geqslant E(X_C)$, iff $B \succeq C$,*
   (v) *if $B \cap C = \varnothing$, then $X_{B \cup C} = X_B + X_C$,*
   (vi) *if $B \neq 0$, then $E(X_B) > 0$,*
   (vii) *$E(X_1^n) = 1$ for every n.*

Moreover, any function ϕ from Re to Re such that $\{\phi(X_B), B \in \mathcal{F}\}$ satisfies (i)–(vii) is the identity function.

58

Proof. First, we have, by familiar arguments from Axioms 1, 3, and 4, the existence of a numerical assignment ϕ. For any B in $\mathcal{F}$, we define the set S of numbers:

(6)
$$S_B = \left\{ \frac{m}{n} : m1 \succcurlyeq nB \right\}.$$

It is easy to show that S is nonempty and has a greatest lower bound, which we use to define ϕ:

$$\phi(B) = g.\ell.b.\ S_B.$$

It is then straightforward to show that, for B and C in $\mathcal{F}$,

(7)
$$\phi(B) \geqslant \phi(C) \text{ iff } B \succcurlyeq C;$$
$$\text{if } B \cap C = \emptyset, \text{ then } \phi(B \cup C) = \phi(B) + \phi(C);$$
$$\text{if } B \neq \emptyset, \text{ then } \phi(B) > 0.$$

Second, it follows from Axiom 2 that

(8)
$$1^n = 1,$$

whence

(9)
$$\phi(1) = 1.$$

From Axiom 6, we infer that, for any object a in G, the numerical sequence

$$1, \phi(a), \phi(a^2), \phi(a^3), \cdots$$

satisfies Inequalities (2) and, hence, determines a unique probability distribution for a, which we represent by the random variable X_a. Furthermore, the expectation function E is defined by $E(X_a^n) = \phi(a^n)$. The independence of mixed moments s and t that have no common object is derived from Axiom 5 by the following argument that uses the sets S_B defined in (6) and their symmetric analogue sets T_B defined below. From 5a, we have at once, if

$$\frac{m}{n} \in S_s \text{ and } \frac{m'}{n'} \in S_t,$$

then

$$\frac{mm'}{nn'} \in S_{st},$$

whence

(10)
$$\phi(s)\phi(t) \geqslant \phi(st).$$

Correspondingly, in order to use Axiom 5b, we define

$$T_B = \left\{ \frac{m}{n} : m1 \preccurlyeq nB \right\}.$$

59

Each set T_B is obviously nonempty and has a least upper bound. We need to show that

$$\ell.u.b.\ T_B = g.\ell.b.\ S_B = \phi(B).$$

Suppose, by way of contradiction, that

$$\ell.u.b.\ T_B < g.\ell.b.\ S_B.$$

(That the weak inequality $\leqslant$ must hold is evident.)
Then there must exist integers m and n such that

$$\ell.u.b.\ T_B < \frac{m}{n} < g.\ell.b.\ S_B.$$

Thus, we have, from the left-hand inequality,

$$m1 \succ nB,$$

and from the right-hand one,

$$m1 \prec nB,$$

which together contradict the weak order properties of $\succcurlyeq$. However, from the definition of T_B, we may then also infer

$$\phi(s)\phi(t) \geqslant \phi(st),$$

which, together with (8), establishes that

$$\phi(s)\phi(t) = \phi(st).$$

The previous argument establishes (i). We next want to show that, with probability one,

$$X_{a\cdot b} = X_a \cdot X_b.$$

We do this by showing the two random variables have identical nth moments for all n. If $a \neq b$, we have independence of X_a and X_b by the argument given previously:

$$E\left(X_{a\cdot b}^n\right) = \phi(a^n \cdot b^n) = \phi(a^n)\phi(b^n) = E\left(X_b^n\right)E\left(X_b^n\right).$$
$$= E\left(X_a^n \cdot X_b^n\right) = E((X_a \cdot X_b)^n),$$

which also establishes (ii) by obvious extension.
If $a = b$, we have the following:

$$E\left(X_{a\cdot a}^n\right) = \phi((a\cdot a)^n) = \phi(a^n \cdot a^n) = \phi(a^{2n})$$
$$= E\left(X_a^{2n}\right) = E\left(X_a^n \cdot X_a^n\right) = E((X_a \cdot X_a)^n),$$

which completes the proof of (iii).

For the empty set, because $\phi(\varnothing) = 0$,

$$X_\varnothing = 0$$

and for B, a nonempty set in $\mathcal{F}$, define

(11) $$X_B = \sum_{s \in B} X_s.$$

Each $s \in B$ is a multinomial moment; thus, X_B is a polynomial in the random variables X_a, with a in some string $s \in B$. Such a random variable is clearly a Borel function, and so its distribution is well defined in terms of the joint product distribution of the independent random variables X_a. Parts (iv)–(vi) of the theorem then follow at once from (7) and (11), and (vii) from (8) and (9).

Finally the uniqueness of the representation follows from the fact that $\phi(\varnothing) = 0$ and $\phi(1) = 1$.

If we specialize the axioms of Definition 3 to qualitative assertions about distributions of a particular form, we can replace Axiom 6 on the complete monotonicity of the sequence of qualitative moments of an object by much simpler conditions. In fact, we know of no simpler qualitative way of characterizing distributions of a given form than by such qualitative axioms on the moments. The following corollary concerns such a characterization of the uniform, binomial, and beta distributions on [0, 1], where the beta distribution is restricted to integer-valued parameters α and β.

COROLLARY TO REPRESENTATION THEOREM *Let* $\mathfrak{A} = (A, \mathcal{F}, G, \succcurlyeq, \cdot, 1)$ *be a structure satisfying Axioms 1–5 of Definition 3, and, for any a in G, assume $a \preccurlyeq 1$.*

I. If the moments of an object a for $n \geqslant 1$ satisfy

$$(n + 1)a^n \sim 2a,$$

then X_a is uniformly distributed on [0, 1].
II. If the moments of an object a for $n \geqslant 1$ satisfy

$$a^n \sim a,$$

then X_a has a Bernoulli distribution on [0, 1].
III. If the moments of an object a for $n \geqslant 1$ satisfy

$$(\alpha + \beta + n)a^{n+1} \sim (\alpha + n)a^n,$$

where α and β are positive integers, then X_a has a beta distribution on [0, 1].

61

Proof. We only give the proof for the Bernoulli distribution in any detail. First, we use the hypothesis $a^n \sim a$ to verify Inequalities (4) and (5). For k even,

$$\left(1 + \binom{k}{2} + \cdots + 1\right)a \succcurlyeq \left(\binom{k}{1} + \cdots + k\right)a$$

certainly holds, and similarly for k odd,

$$\left(1 + \binom{k}{2} + \cdots + k\right)a \succcurlyeq \left(\binom{k}{1} + \cdots + 1\right)a,$$

which shows that Axiom 6 of Definition 3 is satisfied, and so the unique numerical function ϕ of the Representation Theorem exists, with

$$\phi(a) = \phi(a^n) = p$$

for p, some real number such that $0 < p \leqslant 1$, and the distribution is uniquely determined by the moments. The moment-generating function for the Bernoulli distribution with parameter p is, for $-\infty < t < \infty$,

$$\Psi(t) = pe^t,$$

and so the nth derivative of Ψ with respect to t is equal to p at $t = 0$, which completes the proof.

In the case of the beta distribution, we just show how the recursion is derived. Verification of Inequalities (4) and (5) is routine but tedious. The moment-generating function of the beta distribution is not easy to work with, but by direct integration, we have as follows:

$$\begin{aligned} E(X^n) &= \frac{(\alpha + \beta + 1)!}{(\alpha - 1)!(\beta - 1)!} \int_0^1 x^{n+\alpha-1}(1 - x)^{\beta-1}dx \\ &= \frac{(\alpha + \beta - 1)!(\alpha + n - 1)!}{(\alpha + \beta + n - 1)!(\alpha - 1)!} \end{aligned}$$

Using this last expression, it is easy to derive that

$$(\alpha + \beta + n)E(X^{n+1}) = (\alpha + n)E(X^n).$$

Finally the uniform distribution on $[0, 1]$ is a special case of the beta distribution, namely when $\alpha = \beta = 1$.

Note that a Bernoulli distribution of X_a implies that all the probability weight is attached to the end points of the interval, so that, if p is the parameter of the distribution, as in standard notation, then

$$E(X_a) = (1 - p) \cdot 0 + p \cdot 1 = p.$$

We remark informally that some other standard distributions with different domains may also be characterized qualitatively in terms of moments. For

62

example, the normal distribution on $(-\infty, \infty)$ with mean equal to zero and variance equal to one is characterized as follows:

$$a^0 \sim 1,$$
$$a^1 \sim \emptyset,$$
$$a^2 \sim 1,$$
$$a^{2(n+1)} \sim (2n + 1)a^{2n} \text{ for } n \geqslant 1.$$

ACKNOWLEDGMENTS

We are indebted to Duncan Luce for helpful comments on earlier drafts of this chapter. The basic idea of a qualitative theory of moments was presented by the first author some years ago for the special case of the uniform distribution at the 1980 meeting in Madison, Wisconsin, of the Society for Mathematical Psychology.

REFERENCES

Falmagne, J.-C.: 1976, Random conjoint measurement and loudness summation. *Psychological Review*, **83**, 65–79.

Falmagne, J.-C.: 1978, A representation theorem for finite random scale systems. *Journal of Mathematical Psychology*, **18**, 52–72.

Falmagne, J.-C.: 1979, On a class of probabilistic conjoint measurement models: Some diagnostics properties. *Journal of Mathematical Psychology*, **19**, 73–88.

Falmagne, J.-C.: 1980, A probabilistic theory of extensive measurement. *Philosophy of Science*, **47**, 277–296.

Falmagne, J.-C.: 1985, *Elements of psychophysical theory*. New York: Oxford University Press.

Falmagne, J.-C., and Iverson, G.: 1979, Conjoint Weber laws and additivity. *Journal of Mathematical Psychology*, **20**, 164–183.

Gruzewska, H. M.: 1954, L'Arithmétique des variables aléatoires (The arithmetic of random variables). *Cahiers Rhodaniens*, **6**, 9–56.

Hausdorff, F.: 1923, Momentprobleme für ein endliches Intervall (Moment problem for a finite interval). *Mathematische Zeitschrift*, **16**, 220–248.

Iverson, G., and Falmagne, J.-C.: 1985, Statistical issues in measurement. *Mathematical Social Sciences*. **10**, 131–153.

Krantz, D. H., Luce, R. D., Suppes, P. and Tversky, A.: 1971, *Foundations of measurement* (Vol. I). New York: Academic Press.

Shohat, J. A., and Tamarkin, J. D.: 1943, *The problem of moments*. New York: American Mathematical Society.

II

Causality and quantum mechanics

6

Stochastic incompleteness of quantum mechanics

1. INTRODUCTION

The purpose of this article is to bring out in as conceptually clear terms as possible what seems to be a major incompleteness in the probability theory of particles offered by classical quantum mechanics. The exact nature of this incompleteness is illustrated by consideration of some simple quantum-mechanical examples. In addition, these examples are contrasted with the fundamental assumptions of Brownian motion in classical physics on the one hand, and with a controversy of a decade ago in mathematical psychology. (The psychological example is described in detail in the appendix). Our central claim is that classical quantum mechanics is radically incomplete in its probabilistic account of the motion of particles.

In the last part of the article we derive the time-dependent joint distribution of position and momentum of the linear harmonic oscillator, and show how the apparently physically paradoxical statistical independence of position and momentum has a natural explanation. The explanation is given within the framework of the non-quantum-mechanical stochastic theory we construct for such oscillators.

Before entering into any technical details, we shall attempt to spell out in intuitive terms our incompleteness claim. To begin with, there are two senses of completeness that we want to distinguish. One sense of completeness is that demanded by any hidden variable theories of quantum mechanics. States should be found, or be theoretically characterizable at least, that lead to precise values for all observables; that is, no observable is to have a variance greater than zero. This is the sense of completeness that derives from classical deterministic physics; it is not the sense of completeness we are concerned with here.

A second stochastic sense of completeness is that for a family of observables, or a given observable in the more restricted case, all probability questions are

Reprinted from *Synthese* **29** (1974), 311–330.

resolved by the theory. This of course leaves the matter somewhat vague, for even in the most classical settings, for good reason the probability of any subset of the sample space is not considered as having a probability, but only some appropriate σ-algebra of events, and the same is reasonable in quantum mechanics.

A good example of stochastic completeness, but nondeterministic completeness, can be illustrated by the simple case of coin flipping, where the sample space of observables is the set of all possible infinite sequences of heads and tails. Back of the sample space we postulate a state space, with the 'true' probability of the coin coming up heads on a given occasion being a number p between zero and one. A priori we may have some distribution on the unit interval for the true value of p and by experimentation we expect to determine, or to be able to estimate, its true value. Whether we can determine the exact value of p or not, given a value of p, the observables in which we are interested do not become dispersion free, but retain a variance not equal to zero. On the other hand, once the parameter p is fixed or known, all stochastic questions of probability, for example, all joint probabilities over time, are completely determined for the standard σ-algebra of events. This theory, in contrast to the quantum-mechanical examples we shall consider, is stochastically complete.

We also emphasize that we do not have in mind still another discussion of the nonexistence of joint distributions in quantum mechanics, but rather are concerned with the stochastic or temporal character of the probability distributions that can be derived in quantum mechanics and with the extent to which they can be regarded as giving a stochastically complete theory of the motion of a particle.

There is one other point worth clarifying. In the application of continuous-time, continuous-state stochastic processes, it is customary to compute only certain probabilistic quantities and not the full range of what is possible, because of the complexity and difficulty of doing so. The fact that only partial computations are ordinarily made does not disturb in any way our theoretical claim. It still remains pertinent to ask for the characterization of the stochastic process that determines a unique probability measure for any temporal sequence of events we may wish to consider.

It may be responded that one of the characteristic features of quantum mechanics is that proper stochastic processes do not arise. This is exactly the thesis of our article, but we consider it a defect and not a merit of quantum mechanics. In addition, it is our conceptual claim that the difficulties that arise from not having a fully specified stochastic process, even in the simplest cases, provide evidence of the incompleteness of the theory and the extent to which the physical grounds of the theory have not yet been fully thought out. Of course, if one were to take literally a crudely positivistic viewpoint and were prepared only to consider that which is observable, a defense could be made for the incompleteness of quantum mechanics, but we think in practice no one does this. Conceptually it is intuitively impossible to think about quantum-mechanical problems without considering the motion of particles, various constants of the motion, the position of a particle at a given instant, and so forth. Once the motion of particles is discussed and conceptually thought about, it becomes natural to ask for a characterization of the stochastic process that determines a unique measure on the possible sample paths of motion.

Four possibilities remain open. One is that by sufficient effort the quantum-mechanical theory of particles can be made stochastically complete. We conjecture that this is probably not the case, but it is not an easy matter to settle in any definitive way. A second possibility is that although the quantum-mechanical theory is incomplete, it may be completed stochastically in a wide variety of ways that are mathematically consistent with those results that can be derived from quantum mechanics. The third possibility is that quantum mechanics is stochastically incomplete, but that a proper extension to stochastic completeness is impossible and leads to both mathematical and observational inconsistencies. Undoubtedly, many think that this is the case.

Again, the issue is not at all the same as the issue concerning hidden variable theories. We do not believe that this is the situation, and we know of no proofs in the literature that suggest that this is the situation.

The fourth possibility is like the third in that proper extensions lead to mathematical inconsistencies, but different in that stochastically complete theories mathematically inconsistent with quantum mechanics but observationally equivalent to quantum mechanics can be formulated.

Thus, in our view quantum mechanics is stochastically incomplete, but it can, at least in many reasonable cases, be stochastically completed in the sense of either the second or the fourth possibility.

Perhaps more fundamental and important is that qualitative methods of deriving the appropriate differential equations be found if our thesis is correct, and that these methods be used to test the reliability of a stochastic approach to quantum mechanics. The earlier work referred to later in the article has not yet sufficiently explored this avenue in our judgment, and it remains to be seen to what extent a genuine stochastic theory of quantum phenomena beginning from qualitative postulates is possible.

2. TIME-DEPENDENT PROBABILITY DISTRIBUTION
OF A SINGLE OBSERVABLE

To provide a setting for the present discussion, we recall without much exposition some familiar facts; we use the familiar formalism and notation of classical quantum mechanics.

We write the time-dependent Schrödinger equation in the form of Equation (1), and we assume that the wave function Ψ is normalized to one.

$$(1) \qquad \frac{i\hbar \partial \Psi(x,t)}{\partial t} = H\Psi(x,t)$$

The expectation of a given operator A is given by the inner product as expressed in the following equation:

$$(2) \qquad E(A) = (\Psi, A\Psi).$$

If we replace A by $\exp(iuA)$ then we obtain the characteristic function of the probability distribution of the observable A:

$$(3) \qquad \varphi(u) = E(e^{iuA}) = (\Psi, e^{iuA}\Psi).$$

Without entering into details of calculation now (see Sec. 5), we obtain by these methods the time-dependent distribution of an observable for various elementary cases. For example, in the case of a one-dimensional linear harmonic oscillator, we obtain the following probability density for position x at each time t:

$$(4) \qquad f(x,t) = \frac{\alpha}{\pi^{1/2}} e^{-\alpha^2(x - a\cos\omega t)^2}.$$

The various constants occurring in this equation are not of particular interest and are not discussed, except to say that the constant a is related to the averaging over stationary states to obtain the 'mean' probability density. A detailed derivation of (4) is to be found in Schiff (1949, pp. 60–69).

As a different simple example, consider the probability density for the one-dimensional free particle, that is, the particle for which the potential $V = 0$. It may be shown that $f(x,t)$ has the following form:

$$(5) \quad f(x,t) = \left(2\pi\left(\sigma_0^2 + \frac{\hbar^2 t^2}{4m^2\sigma_0^2}\right)\right)^{-1/2} \exp\left(-\frac{x^2}{2(\sigma_0^2 + \hbar^2 t^2/4m^2\sigma_0^2)}\right)$$

where $\sigma_0^2 = \text{var}(X)$ evaluated at $t = 0$. In contrast to the case of the linear harmonic oscillator, the variance of whose distribution is constant, in the case of the free particle, the variance of the distribution, or, as physicists would put it, the uncertainty, increases monotonically in time from $t = 0$ in both past and future directions.

What is important about Equations (4) and (5) is that they represent results that in no way depend on dubious assumptions about how to take the expectation of sums of noncommuting variables or the expectation of the characteristic function of a pair of such variables. They depend only upon a conservative interpretation of the formalism, and one that we believe would be accepted by everyone.

70

We thus have in classical quantum mechanics a general theory of how to get the distribution of position through time, and Equations (4) and (5) are simple examples in two manageable cases. It might therefore seem that the probabilistic theory within quantum mechanics of a single operator or observable is well established and unproblematical in character. We want to show that this is not the case.

The difficulties lie not with equations like (4) and (5), but rather with the fact that they represent the limit in a general way of what can be derived. This means that an underlying stochastic process for a single observable is not determined by the theory, and consequently, a high degree of stochastic incompleteness is a central feature of classical quantum mechanics. One intuitive way of putting the matter is that in equations like (4) and (5) we have the mean distribution through time, but nothing like the characterization of the full process.

To make our central point as forcefully as we can, we adopt three parallel lines of attack. The first is conceptually the simplest; it illustrates the situation by reference to mean learning curves in mathematical psychology. There is an exact parallel to the quantum-mechanical case, but the mathematical formulation is in every respect completely elementary. This argument is given in the appendix. The second line of attack is to contrast the probabilistic results in quantum mechanics with the classical theory of Brownian motion. Finally, the third line of attack, and in certain ways the most interesting, is to construct a classical stochastic theory of the linear harmonic oscillator and to examine its physical interpretation, especially as an extension of the incomplete quantum-mechanical theory.

3. CLASSICAL THEORY OF BROWNIAN MOTION

An explicit and rigorous mathematical theory of Brownian motion has been one of the accomplishments of twentieth-century probability theory. We do not attempt to give a complete formulation, but we can convey the essential conceptual assumptions of the theory. An excellent treatment at an introductory level is found in Karlin (1966). Perhaps the best reference for a thorough study of such processes is Ito and McKean (1965).

Brownian motion as a physical phenomenon was discovered by the English botanist Brown in 1827, and, in one of his most important papers, Einstein in 1905 derived a mathematical description of the phenomenon from basic laws of physics. Let us restrict ourselves to one dimension, and so let $X(t)$ be the random variable at time t whose value is the x component of the particle in Brownian motion. Let x_0 be the position of the particle at time t_0, and let $p(x, t \mid x_0)$ be the conditional probability density of the particle at time $t + t_0$, i.e., the probability density of the random variable $X(t + t_0)$. Einstein argued

71

from physical principles that this conditional density must satisfy the following partial differential equation:

$$(6) \qquad \frac{\partial p(x, t \mid x_0)}{\partial t} = D \frac{\partial^2 p(x, t \mid x_0)}{\partial^2 x},$$

where D is the diffusion coefficient; in the literature this equation is called the diffusion equation. By choosing an appropriate scale, we may take $D = \frac{1}{2}$, and then we can show that the following is the solution of Equation (6):

$$(7) \qquad p(x, t \mid x_0) = \frac{1}{\sqrt{2\pi t}} \exp\left(-\frac{1}{2t}(x - x_0)^2\right),$$

which shows that the conditional density for each time t is a standard Gaussian or normal distribution. Equation (7) is, of course, exactly the sort of equation we obtain in quantum mechanics, e.g., Equations (4) and (5). If this is all that was to be said we would not have established any difference between the classical theory of Brownian motion and quantum mechanics, but of course the probability density function given by (7) is only the beginning of the complete description of the Brownian motion process. The full theory is embodied in the following definition, which we take from Karlin (1966):

DEFINITION *Brownian motion is a stochastic process* $\{X(t); t \geq 0\}$ *with the following properties:*

(a) *Every increment* $X(t + s) - X(s)$ *is normally distributed with mean 0 and variance* ct; $c > 0$ *and c is a constant independent of t;*
(b) *For every pair of disjoint time intervals* $[t_1, t_2]$, $[t_3, t_4]$, *say* $t_1 < t_2 \leq t_3 < t_4$, *the increments* $X(t_4) - X(t_3)$ *and* $X(t_2) - X(t_1)$ *are independent random variables with distributions given in* (a), *and similarly for n disjoint time intervals where n is an arbitrary positive integer.*

From the basic assumptions of this definition, we can derive the joint density f of any finite sequence of random variables $X(t_1), X(t_2), \ldots, X(t_n)$, with $t_1 < t_2 < \cdots < t_n$. In particular, it can easily be shown that

$$f(x_1, \ldots, x_n) = p(x_1, t_1) p(x_2 - x_1, t_2 - t_1)$$
$$(8) \qquad\qquad \cdots p(x_n - x_{n-1}, t_n - t_{n-1}),$$

where

$$(9) \qquad p(x, t) = \frac{1}{\sqrt{2\pi ct}} \exp(-x^2/2ct).$$

It is worth noting that Equation (8) satisfies, in a direct way, the standard conditions that an indexed set of random variables be a stochastic process. Suppose for the present discussion that the indexed set is an interval of real numbers, perhaps the entire set of real numbers T, and for each t in T, there is a random variable $X(t)$. If every finite sequence of random variables indexed on T has a well-defined joint probability distribution, then the entire family of joint distribution functions defines a stochastic process $\{X(t), t \in T\}$.

It is apparent that we are extraordinarily far from realizing this condition for having a stochastic process in any of the standard quantum-mechanical cases in which we have only a mean probability density of the sort exemplified by Equations (4) and (5) for the linear harmonic oscillator and the free particle.

What is important here is not that each well-defined example of quantum mechanics must be extended to become a well-defined stochastic process, but rather the query of what is meant to be the physical interpretation of the probability densities exemplified by Equations (4) and (5). Physicists talk as if the particle has a motion; in fact, we suspect that it would be extraordinarily difficult to eliminate this talk about the motion of a particle from physical discussions of any cases of interest in quantum mechanics, and yet, the theory does not have a natural physical base in terms of the motions of the particles.

We consider this problem a more severe conceptual problem for the clear understanding of quantum mechanics than the general problem of not having joint distributions for noncommuting observables.

Some interesting attempts have been made to develop a stochastic process theory of quantum-mechanical particles as an alternative to standard quantum mechanics. Perhaps the conceptually clearest example of such attempts is provided by Nelson (1967). Nelson explores several simple, but interesting, cases and shows how the Schrödinger equation can be derived from the Markov stochastic process defined for the given physical situation, and correspondingly, how the Markov process can be derived from the Schrödinger equation.

So far as we know, however, there has been no attempt to explore the many different alternatives that are consistent with a given mean probability density, or to examine the physical plausibility of these alternatives.

We now turn to the examination of such an example.

4. DISTINCT MODELS FOR THE SAME MEAN PROBABILITY DENSITY FUNCTION OF THE HARMONIC OSCILLATOR

We first state assumptions that should be satisfied by any stochastic process $\{X(t); -\infty < t < \infty\}$ that has as a consequence the mean Equation (4) for the linear harmonic oscillator.

(i) *The process is Markovian*, i.e., if $X_{t_1}, \ldots, X_{t_n}$ is a finite sequence of random variables with $t_1 < t_2 < \cdots < t_n$, then

$$F\left(X_{t_n} \middle| X_{t_{n-1}}, X_{t_{n-2}}, \ldots, X_{t_1}\right) = F\left(X_{t_n} \middle| X_{t_{n-1}}\right),$$

73

where F is the distribution function. (When the densities exist, we may replace F by f.)

(ii) *The process is Gaussian*, i.e., any finite sequence of random variables as defined in (i) has a multivariate normal (or Gaussian) distribution, and for each t the density $f(x, t)$ is defined by Equation (4).

(iii) *The process is mean-square continuous.*

The concept of continuity introduced in (iii) can usefully be defined in terms of the *correlation function* $\Gamma(t_1, t_2)$, which is itself defined as follows:

$$(10) \qquad \Gamma(t_1, t_2) = E(X(t_1)X(t_2))$$

$$= \int_{-\infty}^{\infty} \int_{-\infty}^{\infty} x_1 x_2 f(x_1, t_1; x_2, t_2) \, dx_1 \, dx_2,$$

when the integrals exist. The stochastic process is mean-square continuous if for every t

$$(11) \qquad \lim_{\tau \to 0, \tau' \to 0} \Gamma(t + \tau, t + \tau') = \Gamma(t, t).$$

Under the strong Gaussian assumption (ii) the process is completely determined by specifying the correlation function $\Gamma(t_1, t_2)$. We have added the Markovian restriction (i) because of its intuitive physical appeal, corresponding as it does to the concept of state in classical mechanics. In similar fashion the continuity requirement is physically natural.

Our fundamental point is that many different models may be chosen that satisfy assumptions (i)–(iii). The choice reduces to a choice of the correlation function $\Gamma(t_1, t_2)$. Moreover, until the correlation function is specified, the probabilistic or stochastic aspects of the theory of the time-dependent behavior of a quantum-mechanical oscillator are incomplete. We emphasize once again that this kind of incompleteness is only uncovered by considering time-dependent phenomena – it does not arise for a fixed Hilbert space at a given time t_0. Thus, in one clear sense, the kind of stochastic incompleteness identified is distinct from that associated with quantum logic in the standard literature.

What is important and surprising, and so far as we know, not previously observed in the literature, is that a highly intuitive and quite simple combination of classical mechanics and the classical theory of stochastic processes may be used to define a stochastic process that has Equation (4) as a consequence.

We first of all recall that the differential equation of the linear harmonic oscillator in classical mechanics is

$$(12) \qquad \ddot{x} + \omega^2 x = 0,$$

with one form of the general solution being

$$(13) \qquad x = a \cos \omega t,$$

74

which is the term that appears in (4). Let X_t now be the random variable whose value is the position of the oscillator at time t, and $\{X_t, -\infty < t < \infty\}$ the corresponding stochastic process. Let Y_t be a random variable with normal or Gaussian distribution having mean zero and variance equal to one, i.e.,

$$(14) \qquad\qquad Y_t \sim N(0, 1).$$

Then from inspection of (4) we see at once that

$$(15) \qquad\qquad X_t = a \cos \omega t + c Y_t,$$

i.e., the position random variable of the oscillator at time t is just the sum of the constant term $a \cos \omega t$ and the 'fluctuation' random variable cY_t that has a normal distribution, where the constant $c = 1/\sqrt{2}\alpha$. The fluctuations Y_t in the motion of the particle play the role that errors of measurement play in the classical theory of observation. It is important to emphasize, however, that we are interpreting Y_t as a physical fluctuation in the motion of the particle not arising from a measurement, but having the same conceptual basis as the fluctuations of Brownian motion. Here, of course, the variance is constant in time and there is no dispersion through time as in the case of Brownian motion.

At first glance it might seem that the stochastic process we propose for the behavior of an oscillator is completely specified by assumptions (i)–(iii) and (14) and (15), but this is not the case. The correlation function Γ is still undetermined. It is easy to show that it must be a symmetric function and in the present case depends only on the difference $\tau = t_2 - t_1$. Because for all t, Y_t has the distribution $N(0, 1)$, the stochastic process is stationary and by assumption (iii) mean-square continuous.

A fundamental theorem of Khinchine's (see Gnedenko, 1968, p. 387) states that for such a continuous stationary process the correlation function $\Gamma(\tau)$ must have the form

$$(16) \qquad\qquad \Gamma(\tau) = \int \cos \tau x \, dF(x),$$

where $F(x)$ is some distribution function, and for any distribution function $F(x)$ there is such a process.

So far as we can see, quantum mechanics imposes no special constraints on $\Gamma(\tau)$, and a wide variety of choices can consistently be made in accordance with (16). This possible variety of correlation functions expresses in a particularly simple way the stochastic incompleteness of quantum mechanics.

The formulation we have given of a stochastic theory for the harmonic oscillator is a direct and, we believe, relatively natural extension of what is to be found in the usual quantum-mechanical discussions of the oscillator. There is, however, another way of looking at the stochastic theory that would provide

75

strong, if not complete, constraints on the correlation function $\Gamma(\tau)$. This is to take as fundamental the energy states of the oscillator. The transition probabilities between these energy states define the fundamental stochastic process, which is now a continuous-time discrete-state process, and the probability distribution of position is given theoretically, conditional on each of the states. As has already been indicated, Equation (4) arises in this theory by averaging over the energy states and thus over the conditional probability distributions. Amusingly enough, the contrast between the theory we have given above and the more extended theory is precisely the contrast that exists in the two forms of learning theory described in the appendix. The linear incremental model is exactly the sort of theory we have defined above, and the all-or-none model with unobservable states of conditioning corresponds to the discrete-state model just sketched.

The central point we have wanted to make is that the standard quantum-mechanical theory of the oscillator is stochastically incomplete, and we have tried to point out as clearly as possible the nature of the incompleteness.

5. PHYSICAL PECULIARITIES OF JOINT DISTRIBUTIONS THAT ARE PROPER

Although noncommuting observables do not for most states ψ have a proper joint distribution, there are many special cases in which they do. We want to examine one such case in some detail – that of the linear harmonic oscillator, where, as before, we 'average' over the states – in order to bring out the central feature of the joint distribution that can be explained by the stochastic theory characterized in the previous part. What we are interested in is once again the time-dependent case. We derive the joint probability density function for position and momentum and show that the distributions of position and momentum are independent through time, which seems physically surprising in terms of ordinary physical ideas of the motion of a particle.

To begin with, it is known from the literature (Schiff, 1949, pp. 60–69) that for the case of a one-dimensional harmonic oscillator

$$\psi(x, t) = \frac{\alpha^{1/2}}{\pi^{1/4}} \exp\left[-\tfrac{1}{2}(\alpha x - \alpha a \cos \omega t)^2 \right.$$
$$\left. - i\left(\tfrac{1}{2}\omega t + \alpha^2 ax \sin \omega t - \tfrac{1}{4}\alpha^2 a^2 \sin 2\omega t\right)\right],$$

where α, a, and ω are physical constants that we need not be concerned with, except that $\alpha^2 = \omega/\hbar$. For convenience of calculation, we replace the momentum p by the propagation 'vector' $k = p/\hbar$. By familiar methods, we may then show, since position and momentum are canonically conjugate operators, that the joint density $f(k, x, t)$ is given by:

$$f(k, x, t) = \frac{1}{\pi} \int \psi^*(x - u/2)e^{-iku}\psi(x + u/2)\,du.$$

Using the above expression for $\psi(x, t)$, we first find that

$$\psi^*(x - u/2, t) = \frac{\alpha^{1/2}}{\pi^{1/4}} \exp\left[-\tfrac{1}{2}(\alpha(x - u/2) - \alpha a \cos \omega t)^2\right.$$
$$\left. + i\left(\tfrac{1}{2}\omega t + \alpha^2 a(x - u/2) \sin \omega t - \tfrac{1}{4}\alpha^2 a^2 \sin 2\omega t\right)\right]$$

and

$$\psi(x + u/2, t) = \frac{\alpha^{1/2}}{\pi^{1/4}} \exp\left[-\tfrac{1}{2}(\alpha(x + u/2) - \alpha a \cos \omega t)^2\right.$$
$$\left. - i\left(\tfrac{1}{2}\omega t + \alpha^2 a(x + u/2) \sin \omega t - \tfrac{1}{4}\alpha^2 a^2 \sin 2\omega t\right)\right].$$

Combining results, integrating, and simplifying, we then obtain

$$f(k, x, t) = \frac{1}{\pi} \exp[-\alpha^2 x^2 - k^2/\alpha^2 - \alpha^2 a^2$$
$$+ 2\alpha^2 xa \cos \omega t - 2ak \sin \omega t],$$

but this joint density may be written in a form that shows directly the statistical independence of k and x by using the fact that $\cos^2 \omega t + \sin^2 \omega t = 1$:

$$f(k, x, t) = \left[\frac{1}{\sqrt{\pi}} \exp{-\alpha^2(x^2 - 2ax \cos \omega t + a^2 \cos^2 \omega t)}\right]$$
$$\cdot \left[\frac{1}{\sqrt{\pi}} \exp{-\left[\frac{k^2}{\alpha^2} + 2ak \sin \omega t + \alpha^2 a^2 \sin^2 \omega t\right]}\right]$$
$$= \left[\frac{\alpha}{\sqrt{\pi}} \exp{-\alpha^2(x - a \cos \omega t)^2}\right]$$
$$\cdot \left[\frac{1}{\alpha\sqrt{\pi}} \exp{-\left(\frac{k}{\alpha} + \alpha a \sin \omega t\right)^2}\right]$$
$$= f(x, t)f(k, t).$$

We note first that as in the case of the position random variable X_t, we can express the momentum – or, more exactly, the propagation – vector K_t as a sum of a constant term $\alpha a \sin \omega t$ and a random variable cZ_t, where Z_t is normally distributed with mean zero and variance 1, i.e.,

$$Z_t \sim N(0, 1)$$

and

$$K_t = -\alpha^2 a \sin \omega t + \frac{\alpha}{\sqrt{2}} Z_t.$$

We also observe that the constant term of K_t is, as would be expected from classical mechanics, the derivative of the constant term of X_t divided by $\hbar$ (and recall that $\alpha^2 = \omega/\hbar$).

77

The statistical independence of position and momentum is now explained simply by assuming that the fluctuations in position are statistically independent of the fluctuations in velocity. In other words, for every t the normally distributed random variables Y_t and Z_t are independent. Our classical stochastic theory of the oscillator thus provides a natural and simple explanation of what might otherwise seem to be a rather puzzling quantum-mechanical result.

6. Concluding Remarks

It might be objected that quantum mechanics should remain in a state of stochastic incompleteness because there is no evidence that a stochastically complete theory of the sort we have developed here or of the sort using diffusion processes (Nelson, 1967) will lead to any new observable phenomena, and thus the stochastic completeness is not only wasted but in a sense misleading. We think that this narrow positivistic view should be rejected. The stochastic theory of quantum phenomena gives a very easily understandable picture of the phenomena, especially, as we have emphasized here, their dynamical aspects. We are not claiming that a classical stochastic theory of all nonrelativistic quantum-mechanical phenomena can be developed, although Bartlett and Moyal (1949) have shown that this is essentially the case when the Hamiltonian is either linear or quadratic as a function of position and momentum. The exact limits of the approach combining classical mechanics and standard stochastic processes need to be examined much more thoroughly than seems yet to have been done.

We do not think it at all out of the question that the stochastic approach could lead to some new predictions that can be experimentally tested.

The second point we want to make is that the stochastic theory we have postulated for the harmonic oscillator is not crypto-deterministic, i.e., the probabilistic fluctuations do not enter only via the initial conditions. Exact knowledge of position of momentum at time t_0, if this were possible, would not lead to exact knowledge of past or future behavior of the oscillator. Even if the energy state could be observed without perturbing the system this indeterminism would still hold. This is for two reasons. First, the transition probabilities from one energy state to another are postulated as fundamentally probabilistic in the more complete stochastic theory already alluded to. Second, the conditional marginal distributions of position and momentum given the energy state are not reducible to deterministic functions, but are as fundamental to a complete classical stochastic theory of the oscillator as they are to the quantum-mechanical theory.

Our third remark concerns our view of the status of quantum logic in a stochastic theory of quantum phenomena. It should be apparent that classical logic untouched and unchanged is the appropriate logic for such stochastic theories. But this does not argue against quantum logic as the logic of observables in quantum experiments. The lattice structure of these observables is a deep-lying fact that is not disturbed by anything we have said.

On the other hand, if it were to turn out in the long run that the classical theory of stochastic processes proved to be the appropriate framework for developing the theory of quantum phenomena in a conceptually natural way, the general significance of quantum logic would almost certainly be reduced.

APPENDIX

Mean learning curves in mathematical psychology

A characteristic controversy in the psychological theory of learning is whether learning takes place on an incremental or an all-or-none basis. From the standpoint of this article, the interesting and subtle point about this controversy is that it is easy to formulate the two extremes of theory so that they both yield the same mean learning curve. The equation for the mean learning curve intuitively corresponds exactly to Equation (4) for the linear harmonic oscillator or to Equation (5) for the free particle. In this discussion and in the development of the other examples, no attempt is made to be mathematically explicit and thereby to make the rigor apparent, but it should be obvious that a mathematically precise formulation of the learning models considered in this part can easily be given.

To keep everything as simple as possible, we assume that exactly two responses are available on each trial. The experimental subject is presented a concept or a stimulus that he must learn to recognize – it does not matter what from the standpoint of theory. One of the responses – let us say response 1 – is the correct response, and the other – response 2 – is incorrect. At the beginning of the experiment the subjects, who are assumed to be homogeneous, will have a probability of responding correctly of $p_{1,1}$. In general, we use the following notation:

$$A_i = \text{response } i, \ i = 1, 2$$
$$P(A_{i,n}) = p_{i,n} = \text{unconditional probability of}$$
$$\text{response } i \text{ on trial } n$$
$$x_n = \text{fixed sequence of responses through trial } n$$
$$P(A_{i,n} \mid x_{n-1}) = p_{i,n}(x) = \text{conditional probability of response}$$
$$i \text{ on trial } n \text{ given prior sequence of}$$
$$\text{responses } x_{n-1}.$$

The all-or-none model has the following simple formulation. An experimental subject begins in an unconditioned state, and on every trial there is a probability

79

c that he will move from the unconditioned to the conditioned state. In addition, the process of moving from the unconditioned to the conditioned state is assumed to be a first-order Markov chain. The transition matrix as indicated has the following form, where U is the unconditioned state, and C is the conditioned state:

	C	U
C	1	0
U	c	$1-c$

The second aspect of the all-or-none model that needs formulation is the probability of a correct response, given the state of conditioning. These assumptions are based on the following two equations:

$$P(A_{1,n} \mid U_n) = p_{1,1},$$
$$P(A_{1,n} \mid C_n) = 1.$$

The critical assumption is that the probability of a correct response, given that the subject stays in the unconditioned state, remains constant, namely, $p_{1,1}$. On the basis of these assumptions, it is easy to derive the mean learning curve for $p_{1,n}$. We first observe that the probability of being in the unconditioned state on trial n is $(1-c)^{n-1}$. Using this result, we then easily obtain the mean learning curve for incorrect responses, i.e., $p_{2,n}$.

$$p_{2,n} = P(A_{2,n} \mid U_n)P(U_n) + P(A_{2,n} \mid C_n)P(C_n)$$
$$= p_{2,1}(1-c)^{n-1} + 0.$$

And thus the mean learning curve for a correct response is:

(17) $$p_{1,n} = 1 - p_{2,1}(1-c)^{n-1}.$$

Let us now look at the linear incremental model that intuitively assumes that the probability of a correct response increases on every trial. The simplest formulation of the model is in terms of a linear transformation on the probability of an incorrect response, given the preceding sequence of responses, where α is the learning parameter that, like c, lies between 0 and 1.

$$p_{2,n}(x) = \alpha p_{2,n-1}(x).$$

From this equation we derive in a direct fashion the mean learning curve for $p_{2,n}$.

$$p_{2,n} = p_{2,1}\alpha^{n-1}.$$

And thus immediately, the mean learning curve for $p_{1,n}$,

(18) $$p_{1,n} = 1 - p_{2,1}\alpha^{n-1}.$$

It is clear that Equations (17) and (18) express exactly the same mean learning curve if we set $\alpha = 1 - c$. Thus from two quite different kinds of assumptions – the assumption that learning occurs suddenly on a single trial, and the incremental assumption that learning increases on each trial, gradually approaching the asymptote of 1 but never achieving it – we get precisely the same mean learning curves. The mean learning curve expressed by either (17) or (18) is the exact correspondence to what we have in Equations (4) and (5) for the linear harmonic oscillator and free particle.

On the other hand, the underlying assumptions of the two distinct models permit us to examine data in a decisive way to determine which is being satisfied. It is not pertinent to enter into details, but a detailed discussion can be found, for example, in Suppes and Ginsberg (1963). We can illustrate matters, however, by considering a single conditional probability. Let us suppose that $m < n$, and we assume that an error occurs on trial n; then we have the following two quite distinct conditional probabilities:

(19)

| | All-or-none | $P(A_{1,m} \mid A_{2,n}) = p_{1,1}$ |

| | Incremental | $P(A_{1,m} \mid A_{2,n}) = 1 - p_{2,1}\alpha^{m-1}$. |

In the case of the all-or-none model, notice that if an error is observed in the protocol of responses on a trial later than m, the implication is that on trial m the subject was in the unconditioned state, and thus, there is no change in the probability of a correct response. In the incremental case the observation of the error on trial n does not influence the gradual increase in the probability of a correct response, and thus, the prediction for trial m is the same whether an error or a correct response occurs on trial n. The simple feature illustrated in these two conditional probabilities is central to disentangling the relative correctness of the two theories with respect to a given set of data.

From a broad conceptual standpoint, it should also be emphasized that the two distinct theories also determine uniquely the probability of any sample path. Contrary to the situation for these learning models, the determination of the sample paths for continuous-time continuous-state processes, the sort most appropriate for most of physics, is a completely nontrivial problem.

In the event of dissatisfaction with the conditioning of the past on the future expressed in Equation (19), a sharp contrast can also be drawn for an arbitrary trial n and the assumption that an error occurred on the previous trial. The all-or-none model 'starts over' at the occurrence of the error, but the incremental model is not affected. This comparison is expressed in the following pair of equations:

| All-or-none | $P(A_{1,n} \mid A_{2,n-1}) = 1 - p_{2,1}(1 - c)$ |

| Incremental | $P(A_{1,n} \mid A_{2,n-1}) = 1 - p_{2,1}\alpha^{n-1}$. |

81

The point being made about the two kinds of learning theories that yield the same mean learning curve can also be developed for continuous-time continuous-state learning processes, but it would take us too deeply into more intricate questions of learning than is desirable. Developments of this kind are given in Suppes and Donio (1967).

One point brought out by the continuous-time learning models that is not transparent in the above formulation concerns the matter of reinforcements. In considering the parallel to quantum mechanics, we might argue that indeed we *can* compute conditional probabilities in quantum mechanics when measurements are taken. Thus, for example, if we make an observation of an observable at a given time t_0, we can then compute the conditional distribution through time of that observable. This point is granted and it is an important part of quantum-mechanical theory, but it is not nearly enough. In the case of the continuous-time learning processes, for example, a continuous sampling of stimuli takes place independent of reinforcements, and it is the responsibility of the theory to deal with the fine structure of this sampling, just as it is the responsibility of a fully formulated, quantum-mechanical dynamical theory to deal with the behavior of the particle through time, independent of measurement interactions. It is not a proper response to say we do not know anything about the particle except when it is being measured, for it is precisely the task of the classical quantum-mechanical theory to compute various constants of the motion, constants that can be computed from the mean curves of the sort expressed by Equations (4) or (5). Furthermore, the standard conceptual way of talking about the motion of a particle implies that the kind of questions we would naturally ask, i.e., questions that correspond to the conditional probabilities in learning theory, can also be answered for quantum mechanics. And the theory should be rich enough in its postulates to answer them.

REFERENCES

Bartlett, M. S., and Moyal, J. E.: 1949, 'The Exact Transition Probabilities of Quantum-mechanical Oscillators Calculated by the Phase-Space Method', *Proceedings of the Cambridge Philosophical Society*, **45**, 545–553.

Gnedenko, B. V.: 1968, *The Theory of Probability* (4th ed.), Chelsea, New York (translated from the Russian).

Ito, K., and McKean, H. P.: 1965, *Diffusion Processes and Their Sample Paths*, Springer, Berlin.

Karlin, S.: 1966, *A First Course in Stochastic Processes*, Academic Press, New York.

Nelson, E.: 1967, *Dynamical Theories of Brownian Motion*, Princeton University Press, Princeton, N. J.

Schiff, L. I.: 1949, *Quantum Mechanics* (1st ed.), McGraw-Hill, New York.

Suppes, P., and Donio, J.: 1967, 'Foundations of Stimulus-sampling Theory for Continuous-Time Processes', *Journal of Mathematical Psychology*, **4**, 202–225.

Suppes, P., and Ginsberg, R.: 1963, 'A Fundamental Property of All-or-None Models, Binomial Distribution of Responses Prior to Conditioning, with Application to Concept Formation in Children', *Psychological Review*, **70**, 139–161.

7

On the determinism of hidden variable theories with strict correlation and conditional statistical independence of observables

1. INTRODUCTION

The main purpose of this chapter is to prove a lemma about random variables, and then to apply this lemma to the characterization of local theories of hidden variables by Bell (1964, 1966) and Wigner (1970), which are focused around Bell's inequality. We use the results of the lemma in two different ways. The first is to show that the assumptions of Bell and Wigner can be weakened to conditional statistical independence rather than conditional determinism because determinism follows from conditional independence and the other assumptions that are made about systems of two spin-$\frac{1}{2}$ particles.

The second direction is to question the attempt of Clauser and Horne (1974) to derive a Bell-type inequality for local stochastic theories of hidden variables which use an assumption of conditional statistical independence for observables.

The main thrust of our analysis obviously arises from the probabilistic lemma we prove. Roughly speaking, this lemma asserts that if two random variables have strict correlation, that is, the absolute value of the correlation is one, and it is in addition assumed that their expectations are conditionally independent when a third random variable λ is given, then the conditional variance of X and Y given λ is zero. In other words, given the hidden variable λ the observables X and Y are strictly determined. The lemma itself, of course, depends on no assumptions about quantum mechanics. It may be regarded as a limitation on any theories that assume both strict correlation between observables and their conditional independence on the basis of some prior or hidden variable.

Reprinted from P. Suppes (Ed.), *Logic and Probability in Quantum Mechanics*. Dordrecht: Reidel, 1976, pp. 445–455.

2. PROBABILISTIC LEMMA ABOUT DETERMINISM

In the statement of the lemma we use standard notation for the expectation (E), covariance (Cov), variance (Var), and standard deviation (σ) of random variables. We use both Var for variance and σ for the standard deviation for compactness of notation.

In most physical discussions of these matters it is assumed that the random variables in question have continuous densities but for the general proof we give here no such assumption is necessary. Finiteness of expectations as indicated in the statement of the lemma is all that is required. The second clause of the lemma just expresses the fact that the correlation is strict, that is, in other notation the absolute value of the correlation $\rho(X, Y) = 1$. The conclusion that the conditional variances of the random variables X and Y are zero is a conclusion that holds with probability one, which is the strongest result of a deterministic kind we would expect in a probabilistic setting.

LEMMA *Let X, Y, and λ be three random variables such that*

(i) $E(XY \mid \lambda) = E(X \mid \lambda)E(Y \mid \lambda)$,
(ii) $|Cov(X, Y)| = \sigma(X)\sigma(Y)$,
(iii) $\sigma(X) > 0$ and $\sigma(Y) > 0$,
(iv) the expectations in (i) and (ii) are finite;

then with probability one

$$\sigma(X \mid \lambda) = \sigma(Y \mid \lambda) = 0.$$

Proof. Note first that by (ii)

(1) $$Y = a + bX \, \text{sign}(Cov(X, Y))$$

with probability one where a, b are real numbers with $b > 0$, and therefore

(2) $$\sigma(Y) = b\sigma(X),$$

(3) $$E(Y \mid \lambda) = a + bE(X \mid \lambda) \, \text{sign}(Cov(X, Y)).$$

Thus, (ii) may be expressed as

(4) $$Cov(X, Y) = b \, Var(X) \, \text{sign}(Cov(X, Y)).$$

We next note the (relatively well-known) fact that

(5) $$Var(X) = E(Var(X \mid \lambda)) + Var(E(X \mid \lambda)).$$

We prove (5) by observing that

(6) $$E(Var(X \mid \lambda)) = E(X^2 \mid \lambda) - E(E^2(X \mid \lambda)),$$

and

(7) $\quad\quad\quad\quad \text{Var}(E(X \mid \lambda)) = E(E^2(X \mid \lambda)) - (E(E(X \mid \lambda)))^2.$

So adding (6) and (7) we get

$$E(E(X^2 \mid \lambda)) - (E(E(X \mid \lambda)))^2 = E(X^2) - E^2(X) = \text{Var}(X).$$

We next observe that

(8) $\quad\quad\quad \text{Cov}(X, Y) = E(E(XY \mid \lambda)) - E(E(X \mid \lambda))E(E(Y \mid \lambda)).$

Combining (8), (i), and (3) we have

$$
\begin{aligned}
(9) \quad \text{Cov}(X, Y) &= E(E(X \mid \lambda)E(Y \mid \lambda)) - E(E(X \mid \lambda))E(E(Y \mid \lambda)) \\
&= E(E(X \mid \lambda)(a + bE(X \mid \lambda) \, \text{sign}(\text{Cov}(X, Y)))) \\
&\quad - E(E(X \mid \lambda))E(a + bE(X \mid \lambda) \, \text{sign}(\text{Cov}(X, Y))) \\
&= b(E(E^2(X \mid \lambda) - E^2(E(X \mid \lambda))) \, \text{sign}(\text{Cov}(X, Y)) \\
&= b \, \text{Var}(E(X \mid \lambda)) \, \text{sign}(\text{Cov}(X, Y)).
\end{aligned}
$$

So by (4) and (5)

$$E(\text{Var}(X \mid \lambda)) = 0,$$

and since $\text{Var}(X \mid \lambda) \geqslant 0$, with probability one $\text{Var}(X \mid \lambda) = 0$. By obvious symmetry of argument, $\text{Var}(Y \mid \lambda) = 0$.

3. Axioms for systems of two spin-$\frac{1}{2}$ particles

Consider a system of two spin-$\frac{1}{2}$ particles initially in the singlet state. Measurements are made of the components of spin for each particle; in particular, let apparatus I measure one particle and apparatus II the other.

There are a number of natural physical assumptions made by Wigner (and Bell at least implicitly), e.g., axial symmetry. These will come out in the axioms. It is to be emphasized that the axioms given here are for this special situation of pairs of spin-$\frac{1}{2}$ particles formed at the source in the singlet spin state with one particle moving to the measuring apparatus I, in one direction, and the other particle to measuring apparatus II in the opposite direction.

The point of the axiomatization is to permit an explicit analysis of just what assumptions are involved in deriving a contradiction between local deterministic theories of hidden variables and the standard quantum mechanical theoretical results for this situation.

Bell states his assumptions basically in terms of expectations; Wigner uses probabilities. Since the random variables whose conditional expectations are the focus of the axioms are two-valued, there is no essential difference between

85

the two approaches. On the other hand, by far the common practice in quantum mechanics is to consider expectations rather than probabilities, and this is the course we have chosen.

Using explicit random-variable notation, the axioms are stated in the spirit of modern probability theory; no additional physical assumptions are left implicit, to be used as needed.

The random variables are these:

ω_I – the direction of orientation of measuring apparatus I;
ω_{II} – the direction of orientation of measuring apparatus II;
M_I – the spin measurement of apparatus I;
M_{II} – the spin measurement of apparatus II;
λ – the hidden variable.

The values of random variables ω_I and ω_{II} are direction vectors, i.e., three-dimensional vectors normed to one, and the cosine of the angle between them is the scalar product $\omega_I \cdot \omega_{II}$, which is itself a new random variable. The values of random variables M_I and M_{II} are $+1$ and -1, for spin $+\frac{1}{2}$ and spin $-\frac{1}{2}$, respectively. Finally, we shall assume for simplicity of notation that λ is a real-valued random variable, but it could be vector-valued and not affect any of the theory, for the essential assumptions about the hidden variable λ are minimal.

As already indicated, an axiom of determinism for the results of a spin measurement given the value of the hidden random variable λ is not required, but can be derived from the weaker axiom of statistical independence.

DEFINITION 1 *A structure* $\langle \omega_I, \omega_{II}, M_I, M_{II}, \lambda \rangle$ *of random variables is a local hidden variable spin-$\frac{1}{2}$ system if and only if the following axioms are satisfied:*

Axiom 1. [Exchangeability] *For any bounded functions f and g the expectations* $E(f(M_I)g(M_{II}) \mid \omega_I, \omega_{II})$ *and* $E(g(M_I)f(M_{II}) \mid \omega_I, \omega_{II})$ *are finite and*

$$E(f(M_I)g(M_{II}) \mid \omega_I, \omega_{II}) = E(g(M_I)f(M_{II}) \mid \omega_I, \omega_{II});$$

Axiom 2. [Axial Symmetry]

(i) $E(M_I \mid \omega_I) = E(M_{II} \mid \omega_{II}) = 0;$

(ii) If we define for any direction vectors ω_I and ω_{II} and any real number α

$$H(\omega_I, \omega_{II}, \alpha) = [\omega_I = \omega_I, \omega_{II} = \omega_{II}, \omega_I \cdot \omega_{II} = \alpha]$$

then

$$E(M_I M_{II} \mid H(\omega_I, \omega_{II}, \alpha)) = E(M_I M_{II} \mid H(\omega_I', \omega_{II}', \alpha')) \text{ if } \alpha = \alpha';$$

Axiom 3. [Opposite Measurement for Same Orientation] *If $\alpha = 1$ then*

$$E(M_I M_{II} \mid H(\omega_I, \omega_{II}, \alpha)) = -1;$$

86

Axiom 4. [Independence of λ] *For all functions g for which the expectations* $\mathrm{E}(g(\lambda))$ *and* $\mathrm{E}(g(\lambda) \mid \omega_\mathrm{I}, \omega_\mathrm{II})$ *are finite,*

$$\mathrm{E}(g(\lambda)) = \mathrm{E}(g(\lambda) \mid \omega_\mathrm{I}, \omega_\mathrm{II});$$

Axiom 5. [Locality: Independence of Orientation of Other Measuring Apparatus]

$$\mathrm{E}(M_\mathrm{I} \mid \omega_\mathrm{I}, \omega_\mathrm{II}, \lambda) = \mathrm{E}(M_\mathrm{I} \mid \omega_\mathrm{I}, \lambda),$$
$$\mathrm{E}(M_\mathrm{II} \mid \omega_\mathrm{I}, \omega_\mathrm{II}, \lambda) = \mathrm{E}(M_\mathrm{II} \mid \omega_\mathrm{II}, \lambda);$$

Axiom 6. [Statistical Independence]

$$\mathrm{E}(M_I M_\mathrm{II} \mid \omega_\mathrm{I}, \omega_\mathrm{II}, \lambda) = \mathrm{E}(M_\mathrm{I} \mid \omega_\mathrm{I}, \omega_\mathrm{II}, \lambda)\mathrm{E}(M_\mathrm{II} \mid \omega_\mathrm{I}, \omega_\mathrm{II}, \lambda).$$

The axioms characterize the probability of a possible outcome of measurements at I and II for two spin-$\frac{1}{2}$ particles σ_1 and σ_2 originally in the singlet state at the source. The first five axioms are implicit in the articles of Bell and Wigner. As already indicated in a general way, Axiom 6 on statistical independence, together with Axioms 1–5, implies deterministic results when the value of λ is given. Axioms 1 and 2 express a broad assumption about symmetry in the experimental measurement procedure.

In the case of Axiom 2, axial symmetry means that specific spatial orientation does not matter, only the angle $\cos^{-1}\alpha$ between the orientation ω_I of apparatus I and the orientation ω_II of apparatus II. Axiom 3 makes explicit a highly specific assumption (or fact perhaps) about spin-$\frac{1}{2}$ particles emitted from a singlet state. If both apparatus I and apparatus II have the same spatial orientation ($\omega_\mathrm{I} \cdot \omega_\mathrm{II} = 1$) then the measurements M_I and M_II must be opposite, i.e., the correlation $\rho(M_\mathrm{I}, M_\mathrm{II}) = -1$. Axioms 1–3 say nothing about hidden variables. Axioms 4–6 do, and the essence of what they postulate is not specific to systems of spin-$\frac{1}{2}$ particles, but rather to local hidden variable theories.

We now prove several simple theorems about the systems defined by Axioms 1–6.

THEOREM 1 [Locality without λ]

$$\mathrm{E}(M_\mathrm{I} \mid \omega_\mathrm{I}, \omega_\mathrm{II}) = \mathrm{E}(M_I \mid \omega_\mathrm{I}),$$
$$\mathrm{E}(M_\mathrm{II} \mid \omega_\mathrm{I}, \omega_\mathrm{II}) = \mathrm{E}(M_\mathrm{II} \mid \omega_\mathrm{II}).$$

Proof.

$$\begin{aligned}
\mathrm{E}(M_\mathrm{I} \mid \omega_\mathrm{I}, \omega_\mathrm{II}) &= \mathrm{E}_\lambda(\mathrm{E}(M_\mathrm{I} \mid \omega_\mathrm{I}, \omega_\mathrm{II}, \lambda)) \\
&= \mathrm{E}_\lambda(\mathrm{E}(M_I \mid \omega_I, \lambda)) \quad \text{by Locality (Axiom 5)} \\
&= \mathrm{E}(M_\mathrm{I} \mid \omega_\mathrm{I}) \quad \text{by Independence of } \lambda \text{ (Axiom 4)}.
\end{aligned}$$

The argument is of course the same for M_II.

THEOREM 2 [Determinism] $\sigma(M_I \mid \omega_I, \lambda) = \sigma(M_{II} \mid \omega_{II}, \lambda) = 0$.

Proof. Immediate from Axioms 3 and 6, and the Probabilistic Lemma of Section 2.

THEOREM 3 [Symmetry]

(i) $\mathrm{Var}(M_I \mid \omega_I) = \mathrm{Var}(M_{II} \mid \omega_{II}) = 1$,
(ii) $\mathrm{Cov}(M_I, M_{II} \mid \omega_I, \omega_{II}) = \mathrm{E}(M_I M_{II} \mid \omega_I, \omega_{II})$.

Proof. Follows directly from Axiom 2 on axial symmetry.

THEOREM 4 *For given direction vectors ω_I and ω_{II}, the joint conditional distribution of random variables M_I and M_{II} is determined by*

$$\mathrm{Cov}(M_I, M_{II} \mid \omega_I = \omega_I, \omega_{II} = \omega_{II}).$$

Proof. Follows from Axiom 2(i), Theorem 1, and Theorem 3.

The intuitive content of Theorem 4 is that the axioms of Definition 1 determine the joint conditional distribution of M_I and M_{II} up to a single parameter, which can be taken to be the conditional covariance.

THEOREM 5 *Let ω_1, ω_2, and ω_3 be three direction vectors, and let Γ be the set of hidden variables, i.e., $\Gamma =$ set of values of the random variable λ. Each λ in Γ belongs to exactly one of the eight regions $\Gamma(\pm, \pm, \pm)$, where*

$$\Gamma(\pm, \pm, \pm) = \{\lambda : \mathrm{E}(M_I \mid \omega_I = \omega_1, \lambda) = \pm 1 \,\&\, \mathrm{E}(M_I \mid \omega_I$$
$$= \omega_2, \lambda) = \pm 1 \,\&\, \mathrm{E}(M_I \mid \omega_I = \omega_3, \lambda) = \pm 1\}.$$

Proof. Follows from Theorem 2 on determinism.

The statement of Theorem 5, but not the notation, makes it clear that the eight regions $\Gamma(\pm, \pm, \pm)$ are relative to the choice of the three direction vectors ω_1, ω_2, and ω_3.

We now derive the basic inequality in terms of covariances.

THEOREM 6 [Basic Inequality] *Let ω_1, ω_2, and ω_3 be three direction vectors and let for $i \neq j$, $1 \leqslant i, j \leqslant 3$,*

$$\mathrm{Cov}(\omega_i, \omega_j) = \mathrm{Cov}(M_I, M_{II} \mid \omega_I = \omega_i, \omega_{II} = \omega_j).$$

Then

$$\mathrm{Cov}(\omega_1, \omega_2) + \mathrm{Cov}(\omega_2, \omega_3) \geqslant \mathrm{Cov}(\omega_3, \omega_1) - 1.$$

Proof. Let $\alpha(+ + +)$ be the probability that λ lies in the region $\Gamma(+ + +)$, with similar notation for the other seven regions.

We note first that

$$\text{Cov}(\omega_i, \omega_j) = \int_\Gamma \text{E}(M_\text{I} M_\text{II} \mid \omega_\text{I} = \omega_i, \omega_\text{II} = \omega_j, \lambda) \, dP(\lambda).$$

Because $\int P(M_\text{I} = 1, M_\text{II} = 1 \mid \omega_\text{I} = \omega_1, \omega_\text{II} = \omega_2, \lambda) \, dP(\lambda) = \alpha(+ - +) + \alpha(+ - -)$, and similarly for other terms, it is easy to show that

$$\text{Cov}(\omega_1, \omega_2) = \alpha(+ - +) + \alpha(+ - -) + \alpha(- + -) + \alpha(- + +)$$
$$- [\alpha(+ + +) + \alpha(+ + -) + \alpha(- - +) + \alpha(- - -)],$$

$$\text{Cov}(\omega_2, \omega_3) = \alpha(+ + -) + \alpha(- + -) + \alpha(+ - +) + \alpha(- - +)$$
$$- [\alpha(+ + +) + \alpha(- + +) + \alpha(+ - -) + \alpha(- - -)],$$

$$\text{Cov}(\omega_3, \omega_1) = \alpha(- + +) + \alpha(- - +) + \alpha(+ + -) + \alpha(+ - -)$$
$$- [\alpha(+ + +) + \alpha(+ - +) + \alpha(- + -) + \alpha(- - -)].$$

Then by direction substitution, and using the fact that the sum of the probability of the eight regions $\Gamma(\pm, \pm, \pm)$ is 1,

$$\text{Cov}(\omega_1, \omega_2) + \text{Cov}(\omega_2, \omega_3)$$
$$= \text{Cov}(\omega_3, \omega_1) - 1 + 4\alpha(+ - +) + 4\alpha(- + -),$$

and since the probabilities $\alpha(+ - +)$ and $\alpha(- + -)$ are nonnegative, the basic inequality follows at once.

The basic inequality of Theorem 6 is slightly different from either one given by Bell or Wigner, because of our explicit use of the standard probabilistic concept of covariance, but the result is really the same, as the next theorem shows.

From the quantum mechanical results given by Bell or Wigner it follows directly that the quantum mechanical covariances are given by the equation:

$$\text{Cov}(\omega_i, \omega_j) = \sin^2 \tfrac{1}{2}\theta_{ij} - \cos^2 \tfrac{1}{2}\theta_{ij},$$

where θ_{ij} is the angle between direction vectors ω_i and ω_j. We are now in a position to obtain the Bell–Wigner contradiction.

THEOREM 7 [Bell–Wigner Contradiction] *The quantum mechanical covariances contradict the basic inequality of Theorem 7 for some directions of measurement ω_1, ω_2, and ω_3.*

Proof. To put the proof exactly in Wigner's form, we note that it follows immediately from the basic inequality of Theorem 7 that we must have

(1) $$\sin^2 \tfrac{1}{2}\theta_{12} + \sin^2 \tfrac{1}{2}\theta_{23} \geqslant \sin^2 \tfrac{1}{2}\theta_{31}.$$

Clearly (1) is violated if ω_2 bisects ω_1 and ω_3, e.g., if $\theta_{12} = \theta_{23} = 30°$ and $\theta_{31} = 60°$.

Violation of (1) is discussed in some detail by Wigner and need not be repeated here. He also points out that (1) implies Bell's original inequality.

4. CAUSALITY AND INDEPENDENCE

In generalizing from deterministic theories of hidden variables it may seem natural to impose a condition of statistical conditional independence, as expressed in the probabilistic lemma of Section 2. A good recent example of the use of such an assumption is to be found in Clauser and Horne (1974). It is our feeling that such an assumption of conditional statistical independence is too strong for a stochastic theory that is not meant to be deterministic. Clauser and Horne do not assume correlations of one between observables – in their case, counts of particles at detectors – and thus our lemma does not apply. But the kind of behavior that must be expected qualitatively can be inferred from assuming that the observables X and Y, as well as the hidden variable λ, are normally distributed. Under the assumption of a multivariate normal distribution for the three random variables, and the assumption that X and Y are conditionally independent, given λ, we can then show that the following relation holds between the correlations whose absolute values are assumed to be strictly between 0 and 1:

$$\rho(X, Y) = \rho(X, \lambda)\rho(Y, \lambda),$$

because the conditional correlation $\rho(X, Y \mid \lambda)$ is given by the same expression as the partial correlation $\rho_{XY \cdot \lambda}$ (for normally distributed random variables)

$$\rho_{XY \cdot \lambda} = \rho(X, Y \mid \lambda) = \frac{\rho(X, Y) - \rho(X, \lambda)\rho(X, \lambda)}{\sqrt{1 - \rho(X, \lambda)^2}.\sqrt{1 - \rho(X, \lambda)^2}}.$$

Assuming symmetry of X and Y, we then get the more restricted expression:

$$\rho(X, Y) = \rho(X, \lambda)^2,$$

which shows that the correlation between X and Y is always strictly less than the correlation between X and λ or between Y and λ. What this relation shows is that if we impose conditional statistical independence then if we have a quite high correlation between observables we must have an even higher correlation between the hidden variable and the observables. Thus, in a clear sense, we must be closer to a deterministic hidden variable theory than to a deterministic theory of the observables.

This admittedly qualitative argument makes us suspicious of the use of an assumption of conditional independence between observables in formulating a hidden variable theory that is meant to be properly stochastic.

Contrary to a standard line of talk about the Einstein–Podolsky–Rosen paradox, in our judgment the absence of conditional statistical independence in a proper stochastic theory of observables does not imply a violation of causality conditions, i.e., does not imply instantaneous action at a distance. A simple classical stochastic model of coin flipping will illustrate the point. Let us assume a hidden variable λ with unknown probability distribution, but let us assume as the distribution of the observable X whose values are heads or tails at detector 1 to be that the probability of heads at detector 1 is λ, the probability therefore of tails is $1 - \lambda$, the probability of heads at detector 2 is $1 - \lambda$, and the probability of tails is λ. Let us also assume that heads has value 1 and tails value -1, and also then a strict correlation of -1 between X and Y. This means that when heads is observed at detector 1 with probability 1, tails is observed at detector 2, and vice versa.

On the other hand, we do not have conditional statistical independence, for it is obvious that it does not hold in the model. In fact, not only is the correlation -1 between X and Y, the conditional correlation, given the value of λ, between X and Y, is -1. In these circumstances, the deterministic results of our probability lemma do not apply and we retain a genuine stochastic process, in particular, $\sigma(X \mid \lambda), \sigma(Y \mid \lambda) > 0$.

At the same time, it is clearly also intuitively wrong to say that because of the conditional correlation of -1 between X and Y, given λ, there is an instantaneous action at a distance between detector 1 and detector 2. The value of λ at the source determines the probabilistic choice of heads or tails for the two detectors and if heads is sent to one detector then tails is sent to the other, just as we might think of a spin-$\frac{1}{2}$ system of particles moving out from the source. What is to arrive at the two detectors is fixed at the source, but fixed in a stochastic fashion. There is no question whatsoever of an instantaneous causal influence between the two sources.

The central point is that when the hidden parameter λ has only a stochastic relationship to the observables, then the absence of conditional statistical independence with respect to λ in no way implies instantaneous action at a distance between the two locations of the detectors or other measuring devices.

REFERENCES

Bell, J. S.: 1964, 'On the Einstein–Podolsky–Rosen Paradox', *Physics*, 1, 195–200.
Bell, J. S.: 1966, 'On the Problem of Hidden Variables in Quantum Mechanics', *Reviews of Modern Physics*, 38, 447–452.
Clauser, J. F., and Horne, M. A.: 1974, 'Experimental Consequences of Objective Local Theories', *Physical Review D*, 10, 526–535.
Wigner, E. P.: 1970, 'On Hidden Variables and Quantum Mechanical Probabilities', *American Journal of Physics*, 38 1005–1009.

8

A new proof of the impossibility of hidden variables using the principles of exchangeability and identity of conditional distributions

The main purpose of this paper is to provide a new proof of the impossibility of local theories of hidden variables based on simpler and more general assumptions than those used by Bell (1964, 1966), Wigner (1970), and Suppes and Zanotti (1976). In particular, no assumptions requiring specific quantum mechanical calculations are required. They are replaced by the principle of exchangeability and the principle of identical conditional distributions given the hidden variable.

The results may be most easily discussed in terms of a system of two spin-$\frac{1}{2}$ particles initially in the singlet state, but generalizations to other quantum mechanical systems of a similar nature are apparent. In essential terms, the analysis of Bell, stated in a particularly clear form by Wigner, depends upon the following assumptions, which we state in intuitive form. (Mathematically explicit axioms are formulated in Appendix A.)

1. *Axial symmetry.* For any direction of the measuring apparatus the expected spin is 0, where spin is measured by $+1$ and -1 for spin $+\frac{1}{2}$ and spin $-\frac{1}{2}$, respectively. Further, the expected product of the spin measurements is the same for different orientations of the measuring apparatuses, as long as the angle between the measuring apparatuses remains the same.

2. *Opposite measurement for same orientation.* The correlation between the spin measurements is -1 if the two measuring apparatuses have the same orientation.

3. *Independence of* λ. The expectation of any function of λ is independent of the orientation of the measuring apparatus.

4. *Locality.* The spin measurement obtained with one apparatus is independent of the orientation of the other measuring apparatus.

5. *Determinism.* Given λ and the orientation of the measuring apparatus, the result of the spin measurement is determined uniquely. In other words, the

Reprinted from P. Suppes (Ed.), *Studies in the Foundations of Quantum Mechanics.* East Lansing, Mich.: Philosophy of Science Association, 1980, pp. 173–191.

conditional variance of the spin measurement, given λ and the direction of the measuring apparatus, is 0.

Suppes and Zanotti (1976) weaken the last assumption of determinism to conditional statistical independence, that is, to the assumption that the expectation of the product of the spin measurements, given λ and the orientation of the measuring apparatuses, is equal to the product of the expectations under the same conditions. The heart of their argument is to show that statistical independence given λ and a correlation of -1 between the measurements implies determinism.

In the present paper we simplify these arguments drastically by using directly de Finetti's principle of exchangeability and the assumption of a negative correlation to prove the nonexistence of a hidden variable theory of the kind that is usually called objective and local. What is significant about our argument, in our judgment, is that it eliminates almost all of the quantum mechanical details from the argument for the nonexistence of hidden variable theories. The conditions that have to be satisfied are quite general. The argument shows how weak the conditions may be that lead to nonexistence of objective local hidden variables.

We need to say something about the principle of exchangeability because it is a principle of symmetry that has not been used in physics, to our knowledge, but has played an important role in subjective theories of probability. The principle was introduced by de Finetti to provide a natural alternative in the subjective theory of probability to objective theories of independence. (We emphasize that the conflicts between objective and subjective theories of probability do not play a role in our own argument here. The principle of exchangeability enters in a formal way and does not require any subjective interpretation, although we are in favor of this broader viewpoint.)

Here is a simple example to illustrate the principle. Suppose we flip 10 times a coin that is known to be biased, and we know artificially that the bias is either .75 in favor of heads or .25 in favor of heads. Then the flips will not be independent because the outcomes of preceding flips will provide information about the probability of a head on the next flip. On the other hand, given the number of heads that occur in 10 flips, the trials on which the heads occur is of no importance. In other words, we have permutational invariance in the sense that the probability of a sequence of 10 outcomes with a fixed number of heads is the same regardless of exactly which trials heads occurred on. Thus, for example, if three heads occurred in 10 flips, then it does not matter whether these three heads occurred on the first three trials, or on the first, seventh, and tenth, etc. Notice how much the principle of exchangeability reduces the number of probabilities that have to be determined. In the present case, instead of considering 2^{10} sequences of possible outcomes, we can reduce this to just

11, the probabilities of 0 to 10 heads, and because of conservation of probability we in fact need to fix only the probability of 10 with the 11th then determined, which is a very considerable reduction of parameters from 2^{10}. The kind of symmetry that is behind the principle of exchangeability and how it applies to the spin experiments should be clear from this example. Exchangeability is at the heart of the symmetry implied by the physics of the spin experiments.

The requirement of exchangeability is uncontroversial. This is not so for the associated requirement of identity of conditional distribution for the two measurements when the apparatuses have the same orientation. No trace of this axiom is to be found in the papers of Bell and Wigner cited above, or in the recent excellent survey paper by Clauser and Shimony (1978), which covers most of the earlier literature. But identity of conditional distribution seems to be a reasonable symmetry requirement, because the hidden variable λ should have the same *expected* causal effect on both particles. Obviously, it is consistent with this axiom that the *actual* causal effects of λ on the particles are different, which is certainly true in the paradigm stochastic experiments on coin flipping.

Especially in relation to the earlier literature on hidden variables, there are two points to be made. As a principle of symmetry, the existence of identical conditional distributions is a different and stronger principle than that of exchangeability – equivalence of the two is true only for infinite sequences of random variables. Second, the principle of identity of conditional distributions is in itself a wholly reasonable principle of symmetry. To take a very elementary example, consider throwing out a pair of fair dice. They have the same conditional distributions on each throw, for there is no principle of difference to make the distributions different, but, of course, we expect the actual outcomes to be different most of the time – five-sixths of the time to be exact. And so it is with spin-$\frac{1}{2}$ particles initially in the singlet state, and a number of other particle systems that are described by Clauser and Shimony as suitable for testing hypotheses about hidden variables.

The fundamental point is that it is the strong symmetry of the principle of identity of conditional distributions that is the main tool for proving the nonexistence of hidden variables.

1. A GENERAL PROBABILISTIC THEOREM ON HIDDEN VARIABLES

In our earlier paper on hidden variable theories (Suppes & Zanotti, 1976), we proved, in a general probabilistic lemma, (i) negative correlation of -1 between two random variables X and Y and (ii) their conditional independence given a random variable λ, that is,

$$E(XY \mid \lambda) = E(X \mid \lambda)E(Y \mid \lambda),$$

imply that λ completely determines the values of X and Y.

In the present context, we establish a necessary and sufficient condition for two-valued exchangeable random variables to have 'identical' causes in the sense of conditional expectation.

THEOREM 1 *Let X and Y be two-valued random variables, for definiteness, with possible values 1 and -1, and with positive variances, i.e., $\sigma(X), \sigma(Y) > 0$. In addition, let X and Y be exchangeable, i.e.,*

$$(1) \qquad P(X = 1, Y = -1) = P(X = -1, Y = 1).$$

Then a necessary and sufficient condition that there exist a hidden variable λ such that $E(XY \mid \lambda = \lambda) = E(X \mid \lambda = \lambda)E(Y \mid \lambda = \lambda)$ and $E(X/\lambda = \lambda) = E(Y \mid \lambda = \lambda)$ for every value λ (except possibly on a set of measure zero) is that the correlation of X and Y be nonnegative.

Proof. We first prove an elementary lemma giving a necessary and sufficient condition for the correlation, $\rho(X, Y)$, of X and Y to be negative.

LEMMA 1 *Under the hypothesis of Theorem 1,*

$$\rho(X, Y) < 0 \quad \text{iff} \quad P(X = 1, Y = -1) > P(X = 1)P(Y = -1).$$

Proof. Since $\sigma(X) = \sigma(Y) > 0$, we have $\rho(X, Y) < 0$ if and only if $\text{cov}(X, Y) < 0$. Next we compute the covariance.

$$\begin{aligned}
\text{cov}(X, Y) &= E(XY) - E(X)E(Y) = E(XY) - E^2(X) \\
&= P(X = 1, Y = 1) + P(X = -1, Y = -1) \\
&\quad - 2P(X = 1, Y = -1) - P^2(X = 1) \\
&\quad - P^2(X = -1) + 2P(X = 1)P(X = -1) \\
&= 4P(X = 1)P(X = -1) - 4P(X = 1, Y = -1).
\end{aligned}$$

Thus, $\text{cov}(X, Y) < 0$ if and only if $P(X = 1, Y = -1) > P(X = 1)P(X = -1)$, as desired.

We next prove half of our theorem, namely, that under the hypotheses stated, conditional independence with identical conditional distributions given a hidden variable λ implies nonnegative correlation.

First, we note, under our assumptions,

$$\begin{aligned}
P(X = 1) = P(Y = 1) &= \int P(X = 1 \mid \lambda = \lambda)\, dP(\lambda) \\
&= \int P(Y = 1 \mid \lambda = \lambda)\, dP(\lambda) \\
&= \pi,
\end{aligned}$$

where, here and in what follows, the integration is from $-\infty$ to ∞. (Actually, it can be shown that for random variables X and Y with only two values, it

95

suffices to suppose λ has only a fixed finite set of values – for either the proof of sufficiency or of necessity.)

Next, we examine

$$P(X = 1, Y = 1) = \int P(X = 1, Y = 1 \mid \lambda = \lambda)\, dP(\lambda)$$

$$= \int P(X = 1 \mid \lambda = \lambda) P(Y = 1 \mid \lambda = \lambda)\, dP(\lambda)$$

$$= \int P^2(X = 1 \mid \lambda = \lambda)\, dP(\lambda),$$

under our hypothesis of identical conditional distributions for X and Y. Now we have from the above by obvious algebraic manipulation

$$P(X{=}1, Y{=}1) = \int (\pi - P(X = 1 \mid \lambda = \lambda))^2\, dP(\lambda)$$

$$- \int \pi^2\, dP(\lambda) + 2 \int \pi P(X = 1 \mid \lambda = \lambda)\, dP(\lambda)$$

$$= \int (\pi - P(X = 1 \mid \lambda = \lambda))^2\, dP(\lambda) + \pi^2$$

$$= \int (\pi - P(X = 1 \mid \lambda = \lambda))^2\, dP(\lambda) + P(X = 1) P(Y = 1),$$

and since the first term is clearly nonnegative, we have

$$P(X = 1, Y = 1) \geqslant P(X = 1) P(Y = 1),$$

and by exactly similar argument

$$P(X = -1, Y = -1) \geqslant P(X = -1) P(Y = -1),$$

whence, adding inequalities and using conservation of probability, we have

$$1 - 2P(X = 1, Y = -1) \geqslant 1 - 2P(X = 1) P(Y = -1),$$

and so

$$P(X = 1, Y = -1) \leqslant P(X = 1) P(Y = -1).$$

Consequently, by Lemma 1,

$$\rho(X, Y) \geqslant 0,$$

as desired.

We now prove that a nonnegative correlation of X and Y implies existence of a hidden variable that will make X and Y be conditionally independent and have identical conditional distributions. We give an elementary proof that such a hidden variable exists. A more general proof using Choquet's theorem is given in Appendix B.

96

Let $\alpha_1 = P(X = 1)$, $\alpha_2 = P(X = 1, Y = 1)$. Then α_1 and α_2 uniquely determine the distribution for the pair of exchangeable random variables X and Y and satisfy conditions:

(i) $\alpha_2 \geqslant 0$;
(ii) $\alpha_1 - \alpha_2 \geqslant 0$;
(iii) $1 - 2\alpha_1 + \alpha_2 \geqslant 0$;
Moreover, $\rho(X, Y) \geqslant 0$ if and only if the following condition is satisfied:
(iv) $\alpha_2 - \alpha_1^2 \geqslant 0$.

Note that condition (iii) is implied by condition (iv).

We now introduce a two-valued random variable λ and show that the conditional probabilities $P(X = 1 \mid \lambda = \lambda_1) = P(Y = 1 \mid \lambda = \lambda_1)$, $P(X = 1 \mid \lambda = \lambda_2) = P(Y = 1 \mid \lambda = \lambda_2)$, and $P(\lambda = \lambda_1)$ may be assigned values compatible with α_1 and α_2.

By hypothesis of conditional independence and identity of conditional distributions for X and Y:

$$
\begin{aligned}
P(X = 1, Y = 1) &= P(X = 1, Y = 1 \mid \lambda = \lambda_1)P(\lambda = \lambda_1) \\
&\quad + P(X = 1, Y = 1 \mid \lambda = \lambda_2)P(\lambda = \lambda_2) \\
&= P^2(X = 1 \mid \lambda = \lambda_1)P(\lambda = \lambda_1) \\
&\quad + P^2(X = 1 \mid \lambda = \lambda_2)P(\lambda = \lambda_2).
\end{aligned}
$$

We now set

$$
P(X = 1 \mid \lambda = \lambda_1) = \frac{1 - \sqrt{1 - 4(\alpha_1 - \alpha_2)}}{2}
$$

$$
P(X = 1 \mid \lambda = \lambda_2) = \frac{1 + \sqrt{1 - 4(\alpha_1 - \alpha_2)}}{2}
$$

$$
P(\lambda = \lambda_1) = \frac{1}{\sqrt{1 - 4(\alpha_1 - \alpha_2)}}\left(\frac{1 + \sqrt{1 - 4(\alpha_1 - \alpha_2)}}{2} - \alpha_1\right).
$$

Then

$$
\begin{aligned}
\alpha_1 = P(X = 1) &= P(X = 1 \mid \lambda = \lambda_1)P(\lambda = \lambda_1) \\
&\quad + P(X = 1 \mid \lambda = \lambda_2)P(\lambda = \lambda_2)
\end{aligned}
$$

and

$$
\begin{aligned}
\alpha_2 = P(X = 1, Y = 1) &= P^2(X = 1 \mid \lambda = \lambda_1)P(\lambda = \lambda_1) \\
&\quad + P^2(X = 1 \mid \lambda = \lambda_2)P(\lambda = \lambda_2),
\end{aligned}
$$

as is easily checked. Moreover, it is also easy to check that

$$
0 \leqslant P(X = 1 \mid \lambda = \lambda_1), P(X = 1 \mid \lambda = \lambda_2), P(\lambda = \lambda_1) \leqslant 1,
$$

and the proof of Theorem 1 is complete.

97

2. Application to Quantum Mechanics

Because the application of hidden variable arguments to a system of two spin-$\frac{1}{2}$ particles initially in the singlet state is now so familiar, and because we have treated it very formally from a probabilistic standpoint in our earlier article (Suppes & Zanotti, 1976), we shall keep the discussion informal here, but it is obvious how it can be made as formal as desired, and this is done in Appendix A.

As we argued in our earlier paper, the notion of locality of an objective hidden variable theory is essentially a principle of conditional statistical independence. Let us now introduce the term *strict local* hidden variable theory for a hidden variable theory that satisfies the following two principles: (i) the principle of conditional statistical independence and (ii) the principle of identity of conditional distributions for particles that satisfy the principle of exchangeability.

Before stating the theorem about strict local hidden variable theories, we want to take note of a point that Clauser and Shimony (1978) make. Although in the Bell arguments it is assumed that when the measuring apparatus is oriented the same in the spin experiments the correlation observed will be -1, in fact, because of the absence of completely efficient measuring apparatus, the actual observed correlation will be definitely negative but not -1. The theorem takes account of this consideration.

THEOREM 2 *Given* (i) *the experimentally determined negative correlation of the spin of two spin-$\frac{1}{2}$ particles that were initially in the singlet state and* (ii) *satisfaction of the principle of exchangeability by the probabilities of spin measurements for the two particles, there can be no strict local hidden variable theory of the phenomena.*

The proof of this theorem follows at once from the hypotheses and Theorem 1.

This negative result raises the question of what kind of theories we are content to have for symmetric phenomena with correlations. What the theorem shows is that no fully satisfactory causal theory is possible for the standard quantum mechanical experiments now discussed extensively in connection with local hidden variable theories.

Let us review the four principles we have been discussing that cannot be simultaneously satisfied.

 (i) Principle of exchangeability: a phenomenological principle of symmetry.
 (ii) Negative correlations between random variables satisfying the principle of exchangeability: a phenomenological result.
 (iii) Identity of conditional distributions given a hidden variable for particles satisfying the principle of exchangeability: a theoretical principle of symmetry.
 (iv) Conditional statistical independence given a hidden variable for particles

98

satisfying the principle of exchangeability: a theoretical principle of locality of causes.

What Theorem 2 shows is that several different kinds of quantum mechanical phenomena cannot simultaneously satisfy all four of these principles, and consequently without any detailed quantum mechanical calculations at all there can be no strict local hidden variable theory.

The question then is what attitude to take about the possibility of theories that are not strict. Much of the physical discussion has argued for the principle of locality as reflected in (iv), but in our view this is *the* principle of the four to be given up, and in our earlier paper we discussed simple examples of stochastic processes that did not satisfy this principle and yet had an apparently satisfactory causal analysis, at least a satisfactory causal analysis up to a certain point – obviously not up to a point that satisfies the principles of conditional independence and of identity of conditional distributions for particles that obey exchangeability.

We modify that simple example of coin flipping so as not to have a perfect negative correlation of -1. Let X be the random variable whose values are heads $(+1)$ or tails (-1) at detector I and let Y be defined similarly at detector II. Then we set

$$P(X = 1, Y = -1 \mid \lambda = \lambda) = P(X = -1, Y = 1 \mid \lambda = \lambda) = \lambda > \tfrac{1}{4}$$

$$P(X = 1, Y = 1 \mid \lambda = \lambda) = P(X = -1, Y = -1 \mid \lambda = \lambda) = \tfrac{1}{2} - \lambda.$$

Note that $\tfrac{1}{4} < \lambda < \tfrac{1}{2}$ by hypothesis, and so $\tfrac{1}{4} < E(\lambda) < \tfrac{1}{2}$. Thus,

$$P(X = 1, Y = -1) = P(X = -1, Y = 1)$$

$$= \int \lambda \, dP(\lambda)$$

$$= E(\lambda)$$

$$P(X = 1, Y = 1) = P(X = -1, Y = -1)$$

$$= \int \left(\tfrac{1}{2} - \lambda \right) dP(\lambda)$$

$$= \tfrac{1}{2} - E(\lambda)$$

$$P(X = 1) = P(Y = 1)$$

$$= \int \tfrac{1}{2} \, dP(\lambda)$$

$$= \tfrac{1}{2}.$$

Consequently,

$$E(X) = E(Y) = 0$$
$$\sigma^2(X) = \sigma^2(Y) = 1,$$

and thus

$$\rho(X, Y) = E(XY) = 1 - 4E(\lambda) < 0.$$

The causal setup is supposed to operate as follows. A device is at work at the source representing the physical embodiment of the hidden variable λ. When λ is the outcome, then the orientation of the coins is determined by a random device using the probability λ according to the equation given above. Use of this random device blocks any further causal anaysis – on our hypothesis of its genuine randomness.

Yet in this example we feel partly satisfied about the causal analysis because the orientation of each coin is fixed by a random device at the source. But we can still feel unsatisfied because of the phenomenological negative correlation. From a causal standpoint we can feel that something is missing that a more complete analysis would permit us to introduce, for example, additional phenomenological variables as well as additional theoretical variables, so that particles put in 'identical circumstances' will, in terms of their probabilistic behavior if not in terms of their behavior on each occasion, not only behave the same but all of the information required to predict their behavior will be embodied in the causal analysis made and, thus, conditional statistical independence will hold. But for such independence to hold we must be able to introduce further variables to eliminate the negative correlations and contravene our hypothesis of genuine randomness.

When we cannot make such improvements, we must face the situation, which we do not always seem willing to do – that for such phenomena we do not have a fully satisfactory causal theory. It seems to us that this is the situation in quantum mechanics. Our general ideas of strict probabilistic causality cannot be satisfied, not to speak of deterministic causation, by quantum mechanics.

Another way of putting it is that when we sacrifice the principle of conditional statistical independence we are saying that we are satisfied with a weaker, that is, a nonstrict, causal theory. No hidden variable theory not satisfying this principle can hope to give us all of the information we would expect to have about the phenomena in question. What is important about the quantum mechanical case is that it seems to be a genuine example of being able to have such a restricted stochastic hidden variable theory. We have termed the principle of conditional statistical independence a principle of locality, but it should be evident that it can be violated by phenomena as in the case of the simple coin-flipping example discussed and yet no hidden action at a distance takes place. The proper

100

interpretation of the locality is that the phenomena cannot be deterministically localized if this principle is not satisfied. It is, of course, an extrapolation to talk about determinism here. As we proved in our previous article, however, in the case of perfect negative correlations, conditional statistical independence does indeed imply determinism, and this, it seems to us, is the reason for baptizing it as a principle of locality.

What remains is the possibility of weak stochastic hidden variable theories satisfying principles (i)–(iii). We have proved that there cannot be strict local hidden variable theories satisfying principles (i)–(iv) for quantum mechanics.

APPENDIX A

Formalization of Theorem 2

We formulated Theorem 2 only informally in the text. Here, in the spirit of our earlier paper, we make the mathematical statement explicit. We change, for the present purpose, the axioms stated in our earlier paper; for example, we now assume only negative correlation, not correlation -1.

The random variables we need are these, stated in terms of the concepts introduced at the beginning of the paper:

ω_I – the direction of orientation of measuring apparatus I;
ω_{II} – the direction of orientation of measuring apparatus II;
M_I – the spin measurement of apparatus I;
M_{II} – the spin measurement of apparatus II;
λ – the hidden variable.

The values of random variables ω_I and ω_{II} are direction vectors, for example, three-dimensional vectors normed to one, and the cosine of the angle between them is the scalar product $\omega_I \cdot \omega_{II}$, which is itself a new random variable. The values of random variables M_I and M_{II} are $+1$ and -1, for spin $+\frac{1}{2}$ and spin $-\frac{1}{2}$, respectively. Finally, we shall assume for simplicity of notation that λ is a real-valued random variable; but it could be vector-valued and not affect any of the theory, for the essential assumptions about the hidden variable λ are minimal.

DEFINITION 1 *A structure* $(\omega_I, \omega_{II}, M_I, M_{II}, \lambda)$ *of random variables is a strict local hidden variable spin-$\frac{1}{2}$ system* *if and only if the following axioms are satisfied:*

Axiom 1. [Exchangeability]

$$E(M_I = 1, M_{II} = -1 \mid \omega_I = \omega_1, \omega_{II} = \omega_2)$$
$$= E(M_I = -1, M_{II} = 1 \mid \omega_I = \omega_1, \omega_{II} = \omega_2).$$

101

Axiom 2. [Axial Symmetry]

(i) $E(M_I \mid \omega_I) = E(M_{II} \mid \omega_{II}) = 0$;
(ii) *If we define for any direction vectors ω_I and ω_{II} and any real number α*

$$H(\omega_I, \omega_{II}, \alpha) = [\omega_I = \omega_I, \omega_{II} = \omega_{II}, \omega_I \cdot \omega_{II} = \alpha],$$

then

$$E(M_I M_{II} \mid H(\omega_I, \omega_{II}, \alpha)) = E(M_I M_{II} \mid H(\omega_I', \omega_{II}', \alpha'))$$

If $\alpha = \alpha'$;

Axiom 3. [Negative Correlation for Same Orientation] If $\alpha = 1$, then

$$\rho(M_I M_{II} \mid H(\omega_I, \omega_{II}, \alpha)) < 0,$$
$$\sigma^2(M_I \mid H(\omega_I, \omega_{II}, \alpha)) > 0,$$
$$\sigma^2(M_{II} \mid H(\omega_I, \omega_{II}, \alpha)) > 0,$$

where ρ is the conditional correlation and σ^2 is the conditional variance.

Axiom 4. [Independence of λ] For all functions g for which the expectations $E(g(\lambda))$ and $E(g(\lambda) \mid \omega_I, \omega_{II})$ are finite,

$$E(g(\lambda)) = E(g(\lambda) \mid \omega_I, \omega_{II}).$$

Axiom 5. [Locality: Independence of Orientation of Other Measuring Apparatus]

$$E(M_I \mid \omega_I, \omega_{II}, \lambda) = E(M_I \mid \omega_I, \lambda),$$
$$E(M_{II} \mid \omega_I, \omega_{II}, \lambda) = E(M_{II} \mid \omega_{II}, \lambda).$$

Axiom 6. [Conditional Statistical Independence]

$$E(M_I M_{II} \mid \omega_I, \omega_{II}, \lambda) = E(M_I \mid \omega_I, \omega_{II}, \lambda) E(M_{II} \mid \omega_I, \omega_{II}, \lambda).$$

Axiom 7. [Identity of Conditional Distributions]

$$E(M_I = 1 \mid \omega_I, \omega_{II}, \lambda) = E(M_{II} = 1 \mid \omega_I, \omega_{II}, \lambda).$$

Using the concepts introduced in Definition 1, we may formulate Theorem 2 more formally as follows.

THEOREM 2 [Formal Version] *There are no strict local hidden variable spin-$\frac{1}{2}$ systems in the sense of Definition 1.*

Proof. In view of Theorem 1, the axioms of Definition 1 are inconsistent, and thus there are no models of the axioms. The proof of inconsistency uses especially – as is obvious from the hypothesis of Theorem 1 – Axioms 1, 3, 6, and 7.

General proof of existence of hidden variables

We give in this Appendix a general proof of the existence of a hidden variable when X and Y have nonnegative correlation and are exchangeable, with no special assumptions about the number of values of λ, as in the elementary proof given in the text for the second half of Theorem 1.

Let M denote the class of all probability measures induced by two exchangeable random variables X and Y on the Cartesian product $\{1, -1\} \times \{1, -1\}$ satisfying the conditions

$$P(X = 1, Y = 1) \geqslant P^2(X = 1) \quad \text{for all } P \in M.$$

Obviously, the set M is convex and compact. We now define: *P is an extreme point of M iff if* P_1, $P_2 \in M$, $0 < \alpha < 1$, $P_1 \neq P_2$, *and* $P = \alpha P_1 + (1 - \alpha)P_2$, *then* $\alpha = 0$ *or* $\alpha = 1$.

LEMMA 2 *Let P be an element of M such that* $P(X = 1, Y = 1) = P^2(X = 1)$. *Then P is an extreme point of M.*

Proof. Suppose not; that is, assume $P = \alpha P_1 + (1 - \alpha)P_2$ with $P_1 \neq P_2 \in M$ and $0 < \alpha < 1$. Then

$$\begin{aligned} P(X = 1, Y = 1) &= \alpha P_1(X = 1, Y = 1) \\ &\quad + (1 - \alpha)P_2(X = 1, Y = 1) \\ &\geqslant \alpha P_1^2(X = 1) + (1 - \alpha)P_2^2(X = 1). \end{aligned}$$

Consider now the random variable

$$Z = H P_1(X = 1) + \bar{H} P_2(X = 1)$$

with $P(H) = \alpha$. Apply Jensen's inequality to $f(Z) = Z^2$ and we have

$$E^2(Z) \leqslant E(Z^2),$$

that is,

$$\begin{aligned} (\alpha P_1(X = 1) &+ (1 - \alpha)P_2(X = 1))^2 \\ &< \alpha P_1^2(X = 1) + (1 - \alpha)P_2^2(X = 1), \end{aligned}$$

since equality is attained if and only if $P_1(X = 1) = P_2(X = 1)$. Combining, we obtain

$$\begin{aligned} P(X = 1, Y = 1) &> (\alpha P_1(X = 1) + (1 - \alpha)P_2(X = 1))^2 \\ &= P^2(X = 1). \end{aligned}$$

Now let $\pi_{\{1,-1\}}$ be the set of all probability measures on $\{1, -1\}$. It is clear that for every $p \in \pi_{\{1,-1\}}$ the product measure $p \times p$ is an extreme point of M. The converse is also true. Namely, we have

$$\text{Extreme } (M) = \{p \times p : p \in \pi_{\{1,-1\}}\}.$$

We now show that each extreme point of M is a product measure. Suppose not; that is, assume there exists P, an extreme point of M, such that $P(X = 1, Y = 1) \neq P^2(X = 1)$. Clearly $0 < P(X = 1) < 1$. Define

$$P_1(X = 1) = \frac{P(X = 1, Y = 1)}{P(X = 1)}$$

and

$$P_2(X = 1) = \frac{P(X = -1, Y = 1)}{P(X = -1)}.$$

Then

$$
\begin{aligned}
P(X = 1) &= P(X = 1, Y = 1) + P(X = -1, Y = 1) \\
&= P_1(X = 1)P(Y = 1) + P_2(X = 1)P(X = -1) \\
&= \alpha P_1(X = 1) + (1 - \alpha)P_2(X = 1),
\end{aligned}
$$

and so P is not an extreme point of M, contrary to our supposition. Applying Choquet's theorem (Meyer, 1966, p. 229), we have:

PROPOSITION 1 *Each exchangeable probability measure P on the Cartesian product $\{1, -1\} \times \{1, -1\}$ satisfying*

$$P(X = 1, Y = 1) \geqslant P(X = 1)^2$$

is of the form

$$P = \int_{[0, 1]} p \times p \, \mathrm{d}Q(p),$$

where Q is the distribution induced on the interval $[0, 1]$ by a hidden variable λ.

REFERENCES

Bell, J. S.: 1964, On the Einstein–Podolsky–Rosen paradox. *Physics*, **1**, 195–200.
Bell, J. S.: 1966, On the problem of hidden variables in quantum mechanics. *Reviews of Modern Physics*, **38**, 447–452.
Clauser, J. F., and Shimony, A.: 1978, Bell's theorem: Experimental tests and implications. *Reports on Progress in Physics*, **41**, 1881–1927.
Meyer, P. A.: 1966, *Probability and potential*. Waltham, Mass.: Blaisdell.
Suppes, P., and Zanotti, M.: 1976, On the determinism of hidden variable theories with strict correlation and conditional statistical independence of observables. In P. Suppes (Ed.), *Logic and probability in quantum mechanics*. Dordrecht, The Netherlands: Reidel [chap. 7, this volume].
Wigner, E. P.: 1970, On hidden variables and quantum mechanical probabilities. *American Journal of Physics*, **38**, 1005–1009.

9

When are probabilistic explanations possible?

The primary criterion of adequacy of a probabilistic causal analysis is that the causal variable should render the simultaneous phenomenological data conditionally independent. The intuition back of this idea is that the common cause of the phenomena should factor out the observed correlations. So we label the principle the *common cause criterion*. If we find that the barometric pressure and temperature are both dropping at the same time, we do not think of one as the cause of the other but look for a common dynamical cause within the physical theory of meteorology. If we find fever and headaches positively correlated, we look for a common disease as the source and do not consider one the cause of the other. But we do not want to suggest that satisfaction of this criterion is the end of the search for causes or probabilistic explanations. It does represent a significant and important milestone in any particular investigation.

Under another banner the search for common causes in quantum mechanics is the search for hidden variables. A hidden variable that satisfies the common cause criterion provides a satisfactory explanation "in classical terms" of the quantum phenomenon. Much of the earlier discussion of hidden variables in quantum mechanics has centered around the search for deterministic underlying processes, but for some time now the literature has also been concerned with the existence of probabilistic hidden variables. It is a striking and important fact that even probabilistic hidden variables do not always exist when certain intuitive criteria are imposed. One of the simplest examples was given by Bell in 1971, who extended his earlier deterministic work to construct an inequality that is a consequence of assuming that two pairs of values of experimental settings in spin-1/2 experiments must violate a necessary consequence of the common cause criterion, that is, the requirement that a hidden variable render the data conditionally independent. It is easy to show that Bell's inequality is a necessary but not sufficient condition for conditional independence. However,

Reprinted from *Synthese* **48** (1981), 191–199.

we shall not pursue further matters involving specific quantum mechanical phenomena in the present context.

Our aims in this short article are more general. First we establish a necessary and sufficient condition for satisfaction of the common cause criterion for events or two-valued random variables. The condition is existence of a joint probability distribution. We then consider the more difficult problem of finding necessary and sufficient conditions for the existence of a joint distribution. We state and prove a general result only for the case of three (two-valued) random variables, but it has as a corollary a pair of new Bell-type inequalities.

The limitation from a scientific standpoint of the first result on satisfaction of the common cause criterion is evident. The mere theoretical existence of a common cause is often of no interest. The point of the theorem is clarification of the general framework of probabilistic analysis. The theorem was partially anticipated by some unpublished work of Arthur Fine on deterministic hidden variables.

The second theorem about the existence of a joint distribution is more directly applicable as a general requirement on data structures, for it is easy to give examples of three random variables for which there can be no joint distribution. Consider the following. Let X, Y, and Z be two-valued random variables taking the values 1 and -1. Moreover, let us restrict the expectation of the three random variables to being zero, that is,

$$E(X) = E(Y) = E(Z) = 0.$$

Now assume that the correlation of X and Y is -1, the correlation of Y and Z is -1, and the correlation of X and Z is -1. It is easy to show that there can be no joint distribution of these three random variables.

THEOREM ON COMMON CAUSES *Let $X_1, \ldots, X_n$ be two-valued random variables. Then a necessary and sufficient condition that there is a random variable λ such that $X_1, \ldots, X_n$ are conditionally independent given λ is that there exists a point probability distribution of $X_1, \ldots, X_n$.*

Proof. The necessity is trivial. By hypothesis

$$P(X_1 = 1, \ldots, X_n = 1 \mid \lambda = \lambda) = \prod_{i=1}^{n} P(X_i = 1 \mid \lambda = \lambda).$$

We now integrate with respect to λ, which has, let us say, measure μ, so we obtain

$$P(X_1 = 1, \ldots, X_n = 1) = \int P(X_1 = 1, \ldots, X_n = 1 \mid \lambda = \lambda) \, d\mu(\lambda).$$

The argument for sufficiency is more complex. To begin with, let Ω be the space on which the joint distribution of $X_1, \ldots, X_n$ is defined. Each X_i generates a partition of Ω:

$$A_i = \{\omega : \omega \in \Omega \ \& \ X_i(\omega) = 1\}$$
$$\bar{A}_i = \{\omega : \omega \in \Omega \ \& \ X_i(\omega) = -1\}.$$

Let $\mathcal{P}$ be the partition that is the common refinement of all these two-element partitions. i.e.,

$$\mathcal{P} = \{A_1 \ldots A_n, A_1 \ldots \bar{A}_n, \ldots, \bar{A}_i \ldots \bar{A}_n\},$$

where juxtaposition denotes intersection. Obviously $\mathcal{P}$ has 2^n elements. For brevity of notation we shall denote the elements of partition $\mathcal{P}$ by C_j, and the indicator function for C_j by C_j^c, i.e.,

$$C_j^c(\omega) = \begin{cases} 1 & \text{if } \omega \in C_j \\ 0 & \text{otherwise.} \end{cases}$$

We now define the desired random variable λ in terms of the C_j^c.

(1) $$\lambda = \sum \alpha_j C_j^c$$

where the α_j are disinct real numbers, i.e., $\alpha_i \neq \alpha_j$ for $i \neq j$. The distribution μ of λ is obviously determined by the joint distribution of the random variables $X_1, \ldots, X_n$.

Using (1), we can now express the conditional expectation of each X_i and of their product given λ.

(2) $$\mathrm{E}(X_i \mid \lambda) = \sum_j \frac{C_j^c}{P(C_j)} \int_{C_j} X_i \, d\mu(\lambda)$$

and

(3) $$\mathrm{E}(X_1 \ldots X_n \mid \lambda) = \sum_j \frac{C_j^c}{P(C_j)} \int_{C_j} X_1 \ldots X_n \, d\mu(\lambda).$$

We need to show that the product of (2) over the X_i's is equal to (3). We first note that in the case of (2) or (3) the integrand, X_i in one case, the product $X_1 \ldots X_n$ in the other, has value 1 or -1. (So λ as constructed is deterministic – a point we comment on later.) Second, the integral over the region C_j is just $P(C_j)$. So we have

(4) $$\mathrm{E}(X_i \mid \lambda) = \sum_j \mathrm{sign}_{C_j}(X_i) C_j^c$$

107

where $\text{sign}_{C_j}(X_i)$ is 1 or -1, as the case may be for X_i over the region C_j. From (4) we then have

(5)
$$\prod_{i=1}^{n} E(X_i \mid \lambda) = \prod_i \sum_j \text{sign}_{C_j}(X_i)C_j^c.$$

Given that the product $C_j^c C_{j'}^c = 0$, if $j \neq j'$, we may interchange product and summation in (5) to obtain

(6)
$$\prod_i E(X_i \mid \lambda) = \sum_j \prod_i \text{sign}_{C_j}(X_i)C_j^c,$$

but by the argument already given the right-hand side of (6) is equal to $E(X_1 \ldots X_n \mid \lambda)$ as desired.

There are several comments we want to make about this theorem and its proof. First, because the random variables X_i are two-valued, it is sufficient just to consider their expectations in analyzing their conditional independence. Second, and more important, the random variable λ constructed in terms of the partition $\mathscr{P}$ yields a deterministic solution. This may be satisfying to some, but it is important to emphasize that the artificial character of λ severely limits its scientific interest. What the theorem does show is that the general structural problem of finding a common cause of a finite collection of events or two-valued random variables has a positive abstract solution. Moreover, extensions to infinite collections of events or continuous random variables are possible but the technical details will not be entered into here. We do emphasize that the necessary inference from conditional independence to a joint distribution does not assume a deterministic causal structure.

The place where the abstract consideration of common causes has been pursued the most vigorously is, of course, in the analysis of the possibility of hidden variables in quantum mechanics. Given the negative results of Bell already mentioned, it is clear how the Theorem on Common Causes must apply: the phenomenological events in question do not have a joint distribution. We are reserving for another occasion the detailed consideration of this point.

Within the present general framework it is important to explore further the existence of nondeterministic common causes. Many important constructive examples of such causes are to be found in many parts of science, but the general theory needs more development. One simple example is given at the end of this article.

We turn now to the second theorem about the existence of a joint distribution for three two-valued random variables, which could be the indicator functions, for example, for three events. We assume the possible values as 1 and -1, and

108

the expectations are zero, so the variances are 1 and the covariances are identical to the correlations.

JOINT DISTRIBUTION THEOREM *Let X, Y, and Z be random variables with possible values 1 and −1, and with*

$$E(X) = E(Y) = E(Z) = 0.$$

Then a necessary and sufficient condition for the existence of a joint probability distribution of the three random variables is that the following two inequalities be satisfied.

$$-1 \leqslant E(XY) + E(YZ) + E(XZ)$$
$$\leqslant 1 + 2 \operatorname{Min}\{E(XY), E(YZ), E(XZ)\}.$$

Proof. We first observe that

$$(1) \qquad\qquad E(XY) = p_{11\cdot} - p_{10\cdot} - p_{01\cdot} + p_{00\cdot},$$

where

$$p_{10} = P(X = 1, Y = -1), \text{ etc.}$$

(We use 0 rather than −1 as a subscript for the −1 value for simplicity of notation. The dot refers to Z.) It follows easily from (1) that

$$(2) \qquad\qquad p_{00\cdot} = p_{11\cdot} = \frac{1}{4} + \frac{E(XY)}{4},$$

and similarly

$$(3) \qquad\qquad p_{0\cdot0} = p_{1\cdot1} = \frac{1}{4} + \frac{E(XZ)}{4},$$

$$(4) \qquad\qquad p_{\cdot00} = p_{\cdot11} = \frac{1}{4} + \frac{E(YZ)}{4},$$

$$(5) \qquad\qquad p_{01\cdot} = p_{10\cdot} = \frac{1}{4} - \frac{E(XY)}{4},$$

$$(6) \qquad\qquad p_{0\cdot1} = p_{1\cdot0} = \frac{1}{4} - \frac{E(XZ)}{4},$$

$$(7) \qquad\qquad p_{\cdot01} = p_{\cdot10} = \frac{1}{4} - \frac{E(YZ)}{4}.$$

Using (2)–(7) we can directly derive the following seven equations for the joint distribution – with p_{111} being treated as a parameter along with E(XY), E(YZ), and E(XZ):

(8)

$$
\begin{cases}
p_{110} = \dfrac{1}{4} + \dfrac{E(XY)}{4} - p_{111} \\[2mm]
p_{101} = \dfrac{1}{4} + \dfrac{E(XZ)}{4} - p_{111} \\[2mm]
p_{011} = \dfrac{1}{4} + \dfrac{E(YZ)}{4} - p_{111} \\[2mm]
p_{100} = p_{111} - \dfrac{E(XY)}{4} - \dfrac{E(XZ)}{4} \\[2mm]
p_{010} = p_{111} - \dfrac{E(XY)}{4} - \dfrac{E(YZ)}{4} \\[2mm]
p_{001} = p_{111} - \dfrac{E(XZ)}{4} - \dfrac{E(YZ)}{4} \\[2mm]
p_{000} = \dfrac{1}{4} + \dfrac{E(XY)}{4} + \dfrac{E(XZ)}{4} + \dfrac{E(YZ)}{4} - p_{111}
\end{cases}
$$

From (8) we derive the following inequalities, where $\alpha = 4p_{111}$:

(9)

$$
\begin{cases}
1 + E(XY) \geqslant \alpha \\
1 + E(XZ) \geqslant \alpha \\
1 + E(YZ) \geqslant \alpha \\
E(YZ) + E(XZ) \leqslant \alpha \\
E(XY) + E(YZ) \leqslant \alpha \\
E(YZ) + E(XZ) \leqslant \alpha \\
1 + E(XY) + E(XZ) + (YZ) \geqslant \alpha
\end{cases}
$$

From the last inequality of (9), we have at once

(10) $\qquad -1 \leqslant E(XY) + E(XZ) + E(YZ),$

because α must be nonnegative. Second, taking the maximum of the fourth, fifth, and sixth inequalities and the minimum of the first, second, and third, and adding Min(E(XY), E(XZ), E(YZ)) to both sides, we obtain

(11) $\quad$ E(XY) + E(XZ) + E(YZ) $\leqslant 1 + 2$ Min$\{$E(XY), E(XZ), E(YZ)$\}$.

Inequalities (10) and (11) represent the desired result. Their necessity, i.e., that they must hold for any joint distribution of X, Y, and Z, is apparent from their derivation.

Sufficiency follows from the following argument. Let

$$C_1 = \text{Max}\{E(XY) + E(XZ), E(XY) + E(YZ), E(XZ) + E(YZ)\},$$
$$C_2 = \text{Min}\{E(XY), E(XZ), E(YZ)\}.$$

It is an immediate consequence of (10) and (11) that

(12) $$C_1 \leqslant 1 + C_2,$$

(13) $$1 + C_1 + C_2 \geqslant 0.$$

Assume now that $C_1 \geqslant 0$.

We may then choose $\alpha = 4p_{111}$ so that

$$\alpha = \beta C_1 + (1 - \beta)(1 + C_2), \quad \text{for } 0 \leqslant \beta \leqslant 1.$$

On the other hand, if $C_1 < 0$, choose α so that

$$\alpha = \beta(1 + C_1 + C_2), \quad \text{for } 0 \leqslant \beta \leqslant 1.$$

It is straightforward to show that for either case of C_1, any choice of β in the closed interval $[0, 1]$ will define an $\alpha/4 = p_{111}$ satisfying the distribution equation (8).

The two theorems we have proved can be combined to give a pair of Bell-type inequalities. Two differences from Bell's 1971 results are significant. First, we give not simply necessary, but necessary and sufficient conditions for existence of a hidden variable. Second, we deal with three rather than four random variables. As would be expected from the proofs of the two theorems, our method of attack is quite different from Bell's.

The corollary is an immediate consequence of the two theorems.

COROLLARY ON HIDDEN VARIABLES *Let X, Y, and Z be random variables with possible values 1 and -1, and with*

$$E(X) = E(Y) = E(Z) = 0.$$

Then a necessary and sufficient condition for the existence of a hidden variable or common cause λ with respect to which the three given random variables are conditionally independent is that the phenomenological correlations satisfy the inequalities

$$-1 \leqslant E(XY) + E(YZ) + E(XZ) \leqslant 1 + 2\,\text{Min}\{E(XY), E(YZ), E(XZ)\}.$$

NONDETERMINISTIC EXAMPLE *The deterministic result of the Theorem on Common Causes can, as already indicated, be misleading. We conclude with a simple but important example that is strictly probabilistic.*

Let X and Y be two random variables that have a bivariate normal distribution with $|\rho(X, Y)| \neq 1$, i.e., the correlation to be factored out by a common

111

cause is nondeterministic, and without loss of generality $E(X) = E(Y) = 0$. *It is a standard result that the partial correlation of X and Y with Z held constant is (for a proof, see Suppes, 1970, p. 116):*

$$\rho(XY \cdot Z) = \frac{\rho(X, Y) - \rho(X, Z)\rho(Y, Z)}{\sqrt{1 - \rho^2(X, Z)}\sqrt{1 - \rho^2(Y, Z)}}.$$

Because a multivariate normal distribution is invariant under an affine transformation, we may take

$$E(Z) = 0,$$
$$E(Z^2) = 1.$$

If $\rho(X, Y) \geqslant 0$, *we set*

$$\rho(X, Z) = \rho(Y, Z) = \sqrt{\rho(X, Y)}.$$

If $\rho(X, Y) < 0$, *we set*

$$\rho(X, Z) = -\rho(Y, Z) = \sqrt{|\rho(X, Y)|}.$$

It is straightforward to check that we now have a proper multivariate normal distribution of X, Y, and Z with

$$\rho(XY \cdot Z) = 0$$

and $\rho(X, Z)$ *and* $\rho(Y, Z)$ *nondeterministic.*

REFERENCES

Bell, J. S.: 1971, 'Introduction to the hidden-variable question', in B. d'Espagnat (ed.), *Foundations of quantum mechanics* (Proceedings of the International School of Physics "Enrico Fermi," Course IL). New York: Academic Press, 171–181.

Suppes, P.: 1970, *A Probabilistic Theory of Causality (Acta Philosophica Fennica, 24)*, Amsterdam: North-Holland.

10

Causality and symmetry

This paper is concerned with inferences from phenomenological variables to hidden causes or hidden variables. A number of theorems of a general sort are stated. The paper concludes with a treatment of Bell's inequalities and their generalization to more than four observables.

1. INTRODUCTION

In this paper we are concerned to present a number of theorems about inferences from phenomenological correlations to hidden causes. In other terms, the theorems are mainly theorems about hidden variables. Most of the proofs will not be given but references will be cited where they may be found.

To emphasize conceptual matters and to keep technical simplicity in the forefront, we consider only two-valued random variables taking the values ± 1. We shall also assume symmetry for these random variables in that their expectations will be zero and thus they will each have a positive variance of one. For emphasis we state:

GENERAL ASSUMPTION *The phenomenological random variables X_1, ..., X_N have possible values ± 1, with means $E(X_i) = 0$, $1 \leqslant i \leqslant N$.*

We also use the notation X, Y, and Z for phenomenological random variables. We use the notation $E(XY)$ for covariance which for these symmetric random variables is also the same as their correlation $\rho(X, Y)$.

The basic meaning of causality that we shall assume is that when two random variables, say X and Y are given, then in order for a hidden variable λ to be labeled a cause, it must render the random variables conditionally independent, that is,

(1) $$E(XY \mid \lambda) = E(X \mid \lambda)E(Y \mid \lambda).$$

Reprinted from S. Diner et al. (Eds.), *The Wave–Particle Dualism*. Dordrecht: Reidel, 1984, pp. 331–340.

113

It is worth noting that although the context for discussion of most of what we have to say is quantum mechanics, the reasoning from effects to causes is classical in science and has been an important explicit methodology since the appearance of Laplace's famous memoir of 1774.

2. Two Deterministic Theorems

We begin with a theorem asserting a deterministic result. It says if two random variables have a strict negative correlation then any cause in the sense of (1) must be deterministic, that is, the conditional variances of the two random variables given the hidden variable λ must be zero. We use the notation $\sigma(X \mid \lambda)$ for the conditional standard deviation of X given λ, and its square is, of course, the conditional variance.

THEOREM 1 (Suppes and Zanotti, 1976) *If*

(i) $E(XY \mid \lambda) = E(X \mid \lambda)E(Y \mid \lambda)$
(ii) $\rho(X, Y) = -1$

then

$$\sigma(X \mid \lambda) = \sigma(Y \mid \lambda) = 0.$$

The second theorem asserts that the only thing required to have a hidden variable for N random variables is that they have a joint probability distribution. This theorem is conceptually important in relation to the long history of hidden-variable theorems in quantum mechanics. For example, in the original proof of Bell's inequalities, Bell assumed a causal hidden variable in the sense of (1) and derived from this assumption his inequalities. What Theorem 2 shows is that the assumption of a hidden variable is not necessary in such discussions – it is sufficient to remain at the phenomenological level. Once we know that there exists a joint probability distribution then there must be a causal hidden variable and in fact this hidden variable may be constructed so as to be deterministic. This theorem shows how fundamental the question of the existence of joint probability distributions is in quantum mechanics. Once again another foundational question, namely, in this case the existence of a hidden variable, in fact reduces to the existence of a joint probability distribution.

On the other hand, it is important to emphasize that the hidden variable constructed in the proof of this theorem is physically very unsatisfactory. This is of course natural when one is concerned to show that no hidden-variable theory of even the weakest sort can be consistent with quantum mechanics, but when one wants a positive causal theory it is important to impose additional constraints on the hidden variable. Some of the theorems we consider later do this.

114

THEOREM 2 (Suppes and Zanotti, 1981) *Given phenomenological random variables $X_1, \ldots, X_N$, then there exists a hidden variable λ such that*

$$E(X_1, \ldots, X_N \mid \lambda) = E(X_1 \mid \lambda) \ldots E(X_N \mid \lambda)$$

if and only if there exists a joint probability distribution of $X_1, \ldots, X_N$. Moreover, λ may be constructed as a deterministic cause, i.e., for $1 \leqslant i \leqslant N$

$$\sigma(X_i \mid \lambda) = 0.$$

3. EXCHANGEABILITY

We now turn to imposing some natural symmetry conditions both at a phenomenological and at a theoretical level. The main principle of symmetry we shall use is that of exchangeability. Two random variables X and Y of the class we are studying are said to be *exchangeable* if the following probabilistic equality is satisfied.

(2) $$P(X = 1, Y = -1) = P(X = -1, Y = 1).$$

The first theorem we state shows that if two random variables are exchangeable at the phenomenological level then there exists a hidden causal variable satisfying the additional restriction that they have the same conditional expectation if and only if their correlation is nonnegative.

THEOREM 3 (Suppes and Zanotti, 1980) *If X and Y are exchangeable, then there exists a hidden variable λ such that*

(i) $E(XY \mid \lambda) = E(X \mid \lambda)E(Y \mid \lambda)$
(ii) $E(X \mid \lambda) = E(Y \mid \lambda)$

if and only if

$$\rho(X, Y) \geqslant 0.$$

There are several useful remarks we can make about this theorem. First, the phenomenological principle of symmetry, namely, the principle of exchangeability, has not been used in physics as explicitly as one might expect. In the context of the kinds of experiments ordinarily used to test hidden-variable theories the requirement of phenomenological exchangeability is uncontroversial. On the other hand, the theoretical requirement of identity of conditional distributions does not have the same status. We emphasize that we refer here to the expected causal effect of λ. Obviously the actual causal effects will in general be quite different. We certainly would concede that in many physical situations this principle may be too strong. The point of our theorems about it is to show that once such a strong theoretical principle of symmetry is required

115

then exchangeable and negatively correlated random variables cannot satisfy this theoretical principle of symmetry.

We now turn to the only theorem whose proof we give in this paper. This is the strengthening of Theorem 3 to show that when the correlations are strictly between zero and one then the causal variable cannot be deterministic. What is important from our conceptual standpoint is to show that the principles of symmetry used in Theorem 3 force us to stochastic hidden variables for correlations that are not deterministic or strictly zero.

THEOREM 4 *Given the conditions of Theorem 3, if* $0 < \rho(X, Y) < 1$, *then* λ *cannot be deterministic, i.e.,* $\sigma(X \mid \lambda), \sigma(Y \mid \lambda) \neq 0$.

Proof. We first observe that under the assumptions we have made:

$$\text{Min}\{P(X = 1, Y = -1), P(X = 1, Y = 1), P(X = -1, Y = -1)\} > 0.$$

Now, let Ω be the probability space on which all random variables are defined. Let $\mathcal{A} = \{A_i\}, 1 \leqslant i \leqslant N$ and $\mathcal{H} = \{H_j\}, 1 \leqslant j \leqslant M$ be two partitions of Ω. We say that $\mathcal{H}$ is a *refinement* of $\mathcal{A}$ *in probability* if and only if for all i's and j's we have:

$$\text{If } P(A_i \cap H_j) > 0 \quad \text{then} \quad P(A_i \cap H_j) = P(H_j).$$

Now let λ be a causal random variable for X and Y in the sense of Theorem 3, and let λ have induced partition $\mathcal{H} = \{H_j\}$, which without loss of generality may be assumed finite. Then λ is deterministic if and only if $\mathcal{H}$ is a refinement in probability of the partition $\mathcal{A} = \{A_i\}$ generated by X and Y, for assume, by way of contradiction that this is not the case. Then there must exist i and j such that $P(A_i \cap H_j) > 0$ and

$$P(A_i \cap H_j) < P(H_j),$$

but then $0 < P(A_i \mid H_j) < 1$.

We next show that if λ is deterministic then $E(X \mid \lambda) \neq E(Y \mid \lambda)$, which will complete the proof.

Let, as before, $\mathcal{H} = \{H_j\}$ be the partition generated by λ. Since we know that

$$\sum_j P(X = 1, Y = -1, H_j) = P(X = 1, Y = -1) > 0,$$

there must be an H_j such that

$$P(X = 1, Y = -1, H_j) > 0,$$

but since λ is deterministic, $\mathcal{H}$ must be a refinement of $\mathcal{A}$ and thus as already proved

$$P(X = 1, Y = -1 \mid H_j) = 1,$$

116

whence

$$P(X = 1, Y = 1 \mid H_j) = 0$$
$$P(X = -1, Y = -1 \mid H_j) = 0$$
$$P(X = -1, Y = -1 \mid H_j) = 0,$$

and consequently we have

(3)
$$P(X = 1 \mid H_j) = P(Y = -1 \mid H_j) = 1$$
$$P(X = -1 \mid H_j) = P(Y = 1 \mid H_j) = 0.$$

Remembering that $E(X \mid \lambda)$ is a function of λ and thus of the partition $\mathcal{H}$, we have from (3) at once that

$$E(X \mid \lambda) \neq E(Y \mid \lambda).$$

4. JOINT DISTRIBUTION

Given the covariances (or correlations) of N random variables of the sort we are considering, it is natural to ask when a compatible joint distribution exists. For $N = 2$, the answer is "always" whatever the correlation, but already for $N = 3$ restrictions are required. For example, if three random variables have identical pairwise correlations of $-\frac{1}{2}$, no compatible joint distribution exists, as is easily checked. A condition for $N = 3$ is this:

THEOREM 5 (Suppes and Zanotti, 1981) *A necessary and sufficient condition for the existence of a joint probability distribution compatible with the given covariances of three phenomenological random variables X, Y, and Z with $E(X) = E(Y) = E(Z) = 0$ is that the following two inequalities be satisfied:*

$$-1 \leqslant E(XY) + E(YZ) + E(XZ) \leqslant 1 + 2 \operatorname{Min}\{E(XY), E(YZ), E(XZ)\}.$$

In view of the following theorem, whose proof is not published as yet, the condition for three is necessary and sufficient for four.

THEOREM 6 *Let $X_1, \ldots, X_N$ phenomenological random variables be given and let N be even. Then a necessary and sufficient condition that there exist a joint probability distribution compatible with the given covariances of all pairs of the phenomenological random variables is that there exist such a compatible distribution for each subset of $N - 1$ variables.*

Finally, we state the only really difficult theorem in this paper. The set of inequalities we give for each N is, we think, about as simple as can be expected purely in terms of the covariances. The inequalities for $N = 3$ given in the theorem are easily shown to be equivalent to the condition of Theorem 5.

117

THEOREM 7 (Suppes and Zanotti, 1991) *A necessary condition that there exist a joint probability distribution compatible with the given covariances of all pairs of N phenomenological random variables is that*

(4) $$\sum_{i<j} a_i a_j E(X_i Y_j) \geqslant \frac{(1-n)}{2}$$

for all subsets of odd cardinality $n \leqslant N$ and with $a_i, a_j = \pm 1$.

(The subscript notation stands for summation over $1 \leqslant i < j \leqslant n$.)

It is easy to show that in the symmetric case of all the covariances being equal, Theorem 7 can be simplified as follows:

THEOREM 8 *If the given covariances of all pairs of phenomenological random variables are equal, then there exists a joint probability distribution compatible with the given covariances if and only if for all $1 \leqslant i < j \leqslant N$*

$$E(X_i X_j) \geq \begin{cases} -\dfrac{1}{N-1} & \text{if } N \text{ is even,} \\ -\dfrac{1}{N} & \text{if } N \text{ is odd.} \end{cases}$$

5. BELL COVARIANCES

First, we recall Bell's inequalities are specifically formulated for measurements of spin of pairs of particles originally in the singlet state. Let A and A' be two possible orientations of apparatus I, and let B and B' be two possible orientations of apparatus II. Let the measurement of spin by either apparatus be 1 or -1, corresponding to spin $\frac{1}{2}$ or $-\frac{1}{2}$, respectively. By $E(AB)$, for example, we mean the expectation of the product of the two measurements of spin, with apparatus I having orientation A and II having orientation B. By axial symmetry, we have $E(A) = E(A') = E(B) = E(B') = 0$, i.e., the expected spin for either apparatus is 0. It is, on the other hand, a well-known result of quantum mechanics that the covariance term $E(AB)$ is:

$$E(AB) = -\cos\theta_{AB},$$

where θ is the difference in angles of orientation A and B. Again, by axial symmetry only the difference in the two orientations matters, not the actual values A and B. (To follow the literature, we begin with the notation A, B, A', and B' for phenomenological random variables, rather than $X_1, \ldots, X_N$, which we go back to later.)

On the assumption that there is a hidden variable that renders the spin results conditionally independent, i.e., that there is a causal hidden variable λ in the sense of equation (1) in the first section, Bell (1964) derives the following inequalities:

$$-2 \leqslant -E(AB) + E(AB') + E(A'B) + E(A'B') \leqslant 2,$$
$$-2 \leqslant E(AB) - E(AB') + E(A'B) + E(A'B') \leqslant 2,$$
(5)
$$-2 \leqslant E(AB) + E(AB') - E(A'B) + E(A'B') \leqslant 2,$$
$$-2 \leqslant E(AB) + E(AB') + E(A'B) - E(A'B') \leqslant 2.$$

This form of the inequalities is due to Clauser, Horne, Shimony and Holt (1969).

We first prove a theorem that uses the condition of Theorem 5 to derive Bell's inequalities. The theorem is essentially equivalent to one stated by Arthur Fine (1982). The proof we give is different from his but not radically so. We include the proof because it is simple and it shows how an elementary, purely phenomenological approach to Bell's inequalities is possible – a fact perhaps not as widely known as it should be.

THEOREM 9 *Bell's inequalities are a necessary consequence of the existence of joint probability distributions for any three of the four phenomenological random variables* $A, A', B,$ *and* B'.

Proof. We apply directly Theorem 5. For each subset of three, $\{A, A', B\}$, $\{A, A', B'\}$, $\{A, B, B'\}$, and $\{A', B, B'\}$ the inequalities of Theorem 5 hold. Adding the four sets of inequalities, and dividing by 2, we have:

(6) $-2 \leqslant E(AA') + E(AB) + E(AB') + E(A'B) + E(A'B') + E(BB') \leqslant 2$
$$+ \text{Min}\{E(AA'), E(AB), E(A'B)\} + \text{Min}\{E(AA'), E(AB'), E(A'B')\}$$
$$+ \text{Min}\{E(AB), E(AB'), E(BB')\} + \text{Min}\{E(A'B), E(A'B'), E(BB')\}$$

To obtain Bell's inequalities from (6), we merely select from each of the four sets that covariance we desire. To exhibit details, we derive the second inequality of (5). Since the right inequality of (6) is satisfied in each case by the minimum of each set, necessarily it is satisfied by any one of the three. Thus we get:

(7) $E(AA') + E(AB) + E(AB') + E(A'B) + E(A'B') + E(BB')$
$$\leqslant 2 + E(AA') + E(AB') + E(A'B) + E(BB').$$

Simplifying (7) we obtain the right-hand side of the second Bell's inequality (5).

(8) $$E(AB) - E(AB') + E(A'B) + E(A'B') \leqslant 2.$$

To derive the left-hand side of this Bell inequality, we use two of the inequalities that follow from Theorem 5. First, for the subset $\{A, A', B\}$, we have

(9) $$-1 \leqslant E(AA') + E(AB) + E(A'B),$$

119

and for the subset $\{A, A', B'\}$

(10) $E(AA') + E(AB') + E(A'B') \leqslant 1 + 2\operatorname{Min}\{E(AA'), E(AB'), E(A'B')\}$
$$\leqslant 1 + 2E(A'B').$$

From (10), we obtain at once by elementary operations:

(11) $\qquad -1 \leqslant -E(AA') - E(AB') + E(A'B').$

Adding now (9) and (11), we get the desired result:

$$-2 \leqslant E(AB) - E(AB') + E(A'B) + E(A'B').$$

The natural converse of Theorem 9 also holds – the proof is due to Arthur Fine.

THEOREM 10 (Fine, 1982) *Bell's inequalities are sufficient for the existence of a joint probability distribution compatible with the given covariances of the phenomenological random variables A, A', B, and B'.*

It is obvious that Theorem 10, not Theorem 9, is physically interesting and rather surprising. Unfortunately the situation becomes more complicated as we go beyond $N = 4$. To examine the general case we need to make explicit the idea of what covariances are given phenomenologically. Garg and Mermin (1982) have given a counterexample to Bell's inequalities being sufficient for eight random variables when what we term *Bell covariances* are given. By Bell covariances we mean covariances $E(X_i X_j)$ for $1 \leqslant i \leqslant i_0 < j \leqslant N$, for some integer i_0. Garg and Mermin's couterexample is for $N = 8$ and $i_0 = 4$. Let $E(X_1 X_5) = E(X_2 X_6) = E(X_3 X_7) = 1$ and $E(X_i X_j) = -\frac{1}{3}$ for $1 \leqslant i \leqslant i_0 < j \leqslant N$. Then it is easy to show that for the quintuple $(X_1, X_3, X_4, X_6, X_8)$ all covariances must be $-\frac{1}{3}$. But it follows at once from Theorem 7 that with $N = 5$, $a_i = a_j = 1$, for existence of a compatible joint distribution $\sum E(X_i X_j) \geqslant -2$, and so there can be no joint distribution compatible with the given covariances all equal to $-\frac{1}{3}$.

We conclude with a probabilistic analysis of Bell covariances. Given such a restricted set of covariances, which arise naturally in quantum mechanics, it is natural to ask under what conditions there exists a compatible joint distribution. We have as an immediate consequence of Theorem 7 the following necessary condition.

THEOREM 11 (Suppes and Zanotti, 1991) *Let Bell covariances $E(X_i X_j)$, $1 \leqslant i \leqslant i_0 < j \leqslant N$, be given. Then a necessary condition that there exist a joint distribution of the N variables compatible with the given covariances is that there exist a solution of the inequalities of Theorem 7, with the non-Bell covariances as unknowns.*

REFERENCES

Bell, J. S.: 1964, On the Einstein–Podolsky–Rosen paradox, *Physics* **1**, pp. 195–200.

Clauser, J. F., Horne, M. A., Shimony, A., and Holt, R. A.: 1969, Proposed experiment to test local hidden-variable theories, *Phys. Rev. Lett.*, **23**, pp. 880–884.

Fine, A.: 1982, Hidden variables, joint probability, and the Bell inequalities, *Phys. Rev. Lett.*, **48**, pp. 291–295.

Garg, A., and Mermin, N. D.: 1982, Correlation inequalities and hidden variables, *Phys. Rev. Lett.*, **49**, pp. 1220–1223.

Laplace, P. S.: 1774, Mémoire sur la probabilité des causes par les événements. Mémoires de l'Académie royale des Sciences de Paris (Savants étrangers), Tome VI, p. 621.

Suppes, P., and Zanotti, M.: 1976, On the determinism of hidden variable theories with strict correlation and conditional statistical independence of observables. In P. Suppes (Ed.), Logic and probability in quantum mechanics. Dordrecht: Reidel, pp. 445–455 [chap. 7, this volume].

Suppes, P., and Zanotti, M.: 1980, A new proof of the impossibility of hidden variables using the principles of exchangeability and identity of conditional distribution. In P. Suppes (Ed.), Studies in the foundations of quantum mechanics. East Lansing, Mich.: Philosophy of Science Association, pp. 173–191 [chap. 8, this volume].

Suppes, P., and Zanotti, M.: 1981, When are probabilistic explanations possible? *Synthese*, **48**, pp. 191–199 [chap. 9, this volume].

Suppes, P., and Zanotti, M.: 1991, New Bell-type inequalities for N > 4 necessary for existence of a hidden variable, *Foundations of Physics Letter*, **4**, pp. 101–107 [chap. 11, this volume].

11

New Bell-type inequalities for $N > 4$ necessary for existence of a hidden variable

1. INTRODUCTION

Fine (1982) proved that Bell's (1964) inequalities are sufficient, as well as necessary, to guarantee the existence of a joint probability distribution of the random variables in question, e.g., those representing certain spin correlation experiments. Garg and Mermin (1982) gave an example of eight random variables satisfying Bell's inequalities but such that no joint distribution exists.

The purpose of this paper is to extend Bell's inequalities to obtain some general necessary conditions for the existence of a joint probability distribution for any finite collection of Bell-type random variables. Our results show that, for $N > 4$, many new elementary inequalities beyond those of Bell must be satisfied by any hidden-variable theory. Since these additional inequalities are violated by quantum mechanical predictions for appropriate choice of measurement arrangements, they serve to increase the conceptual distance between what may be called Einstein locality, after the EPR paradox, and quantum mechanics.

To give a concrete sense of the nature of our results, we exhibit one of the new inequalities violated by Garg and Mermin's example. For theoretical purposes it will suffice to introduce Bell-type random variables $X_1, \ldots, X_N$, $N \geqslant 4$, having possible values ± 1, with means $E(X_i) = 0$ and with what we term *Bell covariances* $E(X_i X_j)$ relative to the index i_0, with $1 \leqslant i \leqslant i_0 < j \leqslant N$, $2 \leqslant i_0 \leqslant N - 2$. Notice that, when $N = 4$, we must, as is familiar, have $i_0 = 2$. Garg and Mermin's counterexample is for $N = 8$ and $i_0 = 4$. Let $E(X_1 X_5) = E(X_2 X_6) = E(X_3 X_7) = 1$ and otherwise $E(X_i X_j) = -\frac{1}{3}$ for $1 \leqslant i \leqslant i_0 < j \leqslant N$. Then it is easy to show that, for the quintuple $(X_1, X_3, X_4, X_6, X_8)$, all covariances must be $-\frac{1}{3}$. No joint distribution can exist for such covariances – a fact that can be computed directly but also follows from Theorem 1 proved below.

Reprinted from *Foundations of Physics Letters*, **4**, no. 1 (1991), 101–107.

As an example, we give a general inequality for $N = 8$ and $i_0 = 4$ which Garg and Mermin's example violates. We use the same five variables they did, but, as will be seen explicitly later, our new Bell-type inequalities use the full set of $N = 8$ variables. Here is the example inequality, where, to simplify notation, we write B_{ij} for $E(X_i X_j)$:

$$-2B_{15} + B_{16} + B_{18} - B_{26} + B_{28} + B_{35} + B_{36} - B_{37}$$
$$+ B_{38} + B_{45} + B_{46} + B_{47} + B_{48} \geqslant -6.$$

It is easy to check that, for the Garg and Mermin example, the right-hand side of the inequality is $-7\frac{1}{3}$, and so it is violated.

We do emphasize that, even for $N = 8$, our full necessary set of new Bell-type inequalities is large, but, as we shall see, the schema representing them is quite simple. In Sec. 2 we prove a general theorem on the existence of a joint probability distribution when *all* covariances are given, not just the Bell ones. This theorem is needed in Sec. 3 when we turn to the main theorem of the paper on Bell-type inequalities for $N > 4$.

Finally, we remark that we conjecture our necessary condition is also sufficient for the existence of a joint probability distribution for $N > 4$, but Pitowsky (to appear) has given a general complexity argument that is a basis for being skeptical of this conjecture.

2. A PRELIMINARY THEOREM

We prove by a simple argument an inequality that follows from the existence of a joint distribution of N random variables. The proof that the inequality for the subsets of odd number is sufficient is much more complicated and is not needed here.

THEOREM 1 *Let* $X_1, \ldots, X_N$ *be* N *random variables with* $X_i = \pm 1$, $E(X_i) = 0$ *and given covariances* $E(X_i X_j)$, $1 \leqslant i < j \leqslant N$. *Then, if a joint distribution of* $X_1, \ldots, X_N$ *exists, we have*

$$(1) \qquad \sum_{i<j} X_i X_j E(X_i X_j) \geqslant \begin{cases} \dfrac{1 - |J|}{2}, & |J| \ odd, \\[2mm] \dfrac{-|J|}{2}, & |J| \ even. \end{cases}$$

for i, j *in* J, *a nonempty subset of* $\{1, \ldots, N\}$ *and* $|J|$ *the cardinality of* J.

Proof. We prove (1) holds for every realization $(x_1, \ldots, x_N)$ of the random variables, where each $x_i = \pm 1$. For $|J|$ odd, since $\sum_{i \in J} x_i X_i$ cannot be zero,

123

we have

$$1 \leqslant E\left(\sum_{i \in J} x_i X_i\right)^2 = E\left(|J| + 2\sum_{i<j} x_i x_j X_i X_j\right)$$

$$= |J| + 2\sum_{i<j} x_i x_j E(X_i X_j),$$

and, in case $|J|$ is even, we have

$$0 \leqslant E\left(\sum_{i \in j} x_i X_i\right)^2.$$

In both cases, (1) is then immediate.

The theorem is easily generalized to random variables which are restricted only to having expectations that are zero and variances that are one, but this generalization is not needed here.

The following corollary whose partial proof from (1) is obvious, will be useful later.

COROLLARY *For $N = 3$, the inequality*

(2) $\qquad X_1 X_2 E(X_1 X_2) + X_1 X_3 E(X_1 X_3) + X_2 X_3 E(X_2 X_3) \geqslant -1$

is necessary and sufficient for existence of a joint probability distribution of $X_1, X_2,$ and X_3 compatible with the given covariances.

Proof. The necessity of (2) follows at once from Theorem 1. Sufficiency can be proved as follows. Suppes and Zanotti (1981) proved that the following inequality is necessary and sufficient:

(3) $\qquad \begin{aligned} -1 &\leqslant E(X_1 X_2) + E(X_1 X_3) + E(X_2 X_3) \\ &\leqslant 1 + 2\min\{E(X_1 X_2), E(X_1 X_3), E(X_2 X_3)\}, \end{aligned}$

and it is easy to show that (2) implies (3).

3. NEW BELL-TYPE INEQUALITIES

We define a *Bell triple* to be three random variables such that exactly the first two have indices either less or greater than i_0, i.e., $1 \leqslant i < j \leqslant i_0 < k \leqslant N$ or $1 \leqslant k \leqslant i_0 < i < j \leqslant N$. For each Bell triple (X_i, X_j, X_k), we define Bell functions B_{ij}:

$$B_{ij}(x_i x_j) = \begin{cases} 1 - \max_k |E(X_i X_k) - x_i x_j E(X_j X_k)|, \\ \text{if } (X_i, X_j, X_k) \text{ is a Bell triple}, \\ x_i x_j E(X_i X_j), \qquad \text{if } 1 \leqslant i \leqslant i_0 < j \leqslant N, \end{cases}$$

124

where $x_i, x_j = \pm 1$. Intuitively the Bell functions provide upper bounds on the non-Bell covariances, but, of course, in the case of the Bell covariances, the Bell functions for given indices are just identical to them, modulo the sign of $x_i x_j$, as can be seen from the second branch of the defining equation. The rationale of the Bell functions becomes clear in the proof of the next theorem, the main one in this paper.

THEOREM 2 *Let $E(X_i X_j)$ be given Bell covariances relative to the index i_0, $1 \leqslant i \leqslant i_0 < j \leqslant N$, for random variables $X_1, \ldots, X_N$. If there exists a joint distribution of $X_1, \ldots, X_N$ compatible with the given Bell covariances, then, for every nonempty subset J of $\{1, \ldots, N\}$, we have*

$$(4) \qquad \sum_{i<j} B_{ij}(x_i x_j) \geqslant \begin{cases} \dfrac{1 - |J|}{2}, & if \ |J| \, odd, \\[2ex] \dfrac{-|J|}{2}, & if \ |J| \, even. \end{cases}$$

for $i, j \in J, x_i, x_j = \pm 1$.

Proof. If a joint distribution for the random variables $X_1, \ldots, X_N$ compatible with the given Bell covariances relative to index i_0 exists, then there exists a joint distribution for each Bell triple (X_i, X_j, X_k) with $E(X_i X_j)$ independent of the choice of X_k. Following the Corollary of Theorem 1, we then have, for any realization (x_i, x_j, x_k),

$$x_i x_j E(X_i X_j) + x_i x_k E(X_i X_k) + x_j x_k E(X_j X_k) \geqslant -1.$$

So, in particular, if $x_i = x_j = 1$, then

$$(5) \qquad E(X_i X_j) + x_k E(X_i X_k) + x_k E(X_j X_k) \geqslant -1;$$

and, if $x_i = 1, x_j = -1$, then

$$(6) \qquad -E(X_i X_j) + x_k E(X_i X_k) - x_k E(X_j X_k) \geqslant -1.$$

From (5) and (6) we infer

$$-1 - x_k(E(X_i X_k) + E(X_j X_k)) \leqslant E(X_i X_j)$$
$$\leqslant 1 + x_k(E(X_i X_k) - E(X_j X_k)),$$

for every Bell triple (X_i, X_j, X_k) and every $x_k = \pm 1$. Thus we have at once

$$-1 + |E(X_i X_k) + E(X_j X_k)| \leqslant E(X_i X_j)$$
$$(7) \qquad\qquad\qquad \leqslant 1 - |E(X_i X_k) - E(X_j X_k)|,$$

and so

$$-1 + \max_k |E(X_i X_k) + E(X_j X_k)| \leqslant E(X_i X_j)$$
$$\leqslant 1 - \max_k |E(X_i X_k) - E(X_j X_k)|.$$

If $x_i x_j = 1$, we have

(8) $$1 - \max_k |E(X_i X_k) - E(X_j X_k)| \geq x_i x_j E(X_i X_j);$$

and, if $x_i x_j = -1$,

(9) $$1 - \max_k |E(X_i X_k) + E(X_j X_k)| \geq x_i x_j E(X_i X_j).$$

Combining (8) and (9), we have for the Bell functions

(10) $$B_{ij}(x_i x_j) = 1 - \max_k |E(X_i X_k) - x_i x_j E(X_j X_k)|$$
$$\geq x_i x_j E(X_i X_j).$$

The desired inequalities (4) follow at once from (10) and (1) of Theorem 1.

COROLLARY *If there exists a hidden variable λ such that $X_1, \ldots, X_N$ are conditionally independent, given λ, i.e.,*

$$E(X_1, \ldots, X_N | \lambda) = E(X_1 | \lambda), \ldots, E(X_N | \lambda),$$

then inequalities (4) must be satisfied.

Proof. Suppes and Zanotti (1981) proved that the assumption of a hidden variable as just formulated is equivalent to the assumption of a joint probability distribution of $X_1, \ldots, X_N$. The corollary follows at once from the theorem and this equivalence.

To show how inequalities (4) work in a particular case, we derive the inequality stated at the beginning which rules out Garg and Mermin's counterexample. Remember $N = 8$, $i_0 = 4$, $J = \{1, 3, 4, 6, 8\}$, $B_{15} = B_{26} = B_{37} = 1$, and $B_{16} = B_{18} = B_{36} = B_{38} = B_{46} = B_{48} = -\frac{1}{3}$.

We need to compute the Bell functions. As a simple case, we take $x_i = x_j = 1$. So we need only to compute B_{13}, B_{14}, B_{34}, and B_{68}. We give details only for B_{13}:

$$B_{13} = 1 - \max_k |E(X_1 X_k) - E(X_3 X_k)|$$
$$= 1 - E(X_1 X_5) + E(X_3 X_5) = -\frac{1}{3},$$

and, similarly,

$$B_{14} = 1 - E(X_1 X_5) + E(X_4 X_5) = -\frac{1}{3},$$
$$B_{34} = 1 - E(X_3 X_7) + E(X_4 X_7) = -\frac{1}{3},$$
$$B_{68} = 1 - E(X_2 X_6) + E(X_2 X_8) = -\frac{1}{3}.$$

The full inequality in terms of Bell covariances is then as given in the beginning of the paper.

126

We can also easily derive the Bell inequality for $N = 4$. This can be done most directly by using (7) of the proof, from which we get at once for a Bell quadruple (X_i, X_j, X_k, X_l), with $1 \leqslant i < j \leqslant i_0 < k < l \leqslant N$ or $1 \leqslant k < l \leqslant i_0 < i < j \leqslant N$,

(11) $\qquad 1 - |E(X_i X_k) - x_i x_j E(X_j X_k)| \geqslant x_i x_j E(X_i X_j)$

and

(12) $\qquad 1 - |E(X_i X_l) - x_i x_j E(X_j X_l)| \geqslant x_i x_j E(X_i X_j).$

Setting $x_i x_j = 1$ in (11) and $x_i x_j = -1$ in (12) and adding the two inequalities, we get

$$|E(X_i X_k) - E(X_j X_k)| + |E(X_i X_l) + E(X_j X_l)| \leqslant 2,$$

a standard form of the Bell inequalities, with obvious permutation of the indices k and l permitted.

REFERENCES

Bell, J. S.: 1964. On the Einstein–Podolsky–Rosen paradox. *Physics*, **1**, 195–200.
Fine, A.: 1982. Hidden variables, joint probability, and the Bell inequalities. *Phys. Rev. Lett*, **48**, 291–295.
Garg, A., & Mermin, N. D.: 1982. Correlation inequalities and hidden variables. *Phys. Rev. Lett*, **49**, 1220–1223.
Pitowsky, I. (to appear).
Suppes, P., & Zanotti, M.: 1981. When are probabilistic explanations possible? *Synthese*, **48**, 191–199 [chap. 9, this volume].

12

Existence of hidden variables having only upper probabilities

Great attention has been given to the exact formulation of locality conditions in the existence of hidden variables for quantum mechanical configurations that do not satisfy Bell's inequalities. In contrast, almost no attention has been given to examining what happens if the requirements for a probability measure are relaxed. The purpose of this paper is to show that local hidden variables do exist if the requirement of a probability measure is weakened to that of having only an upper probability measure. In the past several decades, upper and lower measures have been brought into the theory of statistical inference and the measurement of beliefs rather extensively, but in almost all cases considered in statistical theory or the belief framework, it is assumed that a probability measure exists, or in many cases, that an entire family of probability measures is given such that the sup and inf of the family defines the upper and lower measures. Sometimes still more is required, namely, that the upper–lower measure be a capacity (in the sense of Choquet) of infinite order. This is, for example, true of Dempster's theory of statistical inference. For an extensive recent analysis of the various theories of imprecise statistical models or belief functions using upper and lower probabilities, Walley (1991) is an excellent reference.

In contrast and as is well known, in standard quantum mechanical cases the Bell inequalities cannot be satisfied and this implies that there exists no probability measure. As far as we know, the present application is almost unique in being an example in which upper probabilities can be defined in a natural way and yet there is no physically possible probability measure. Why we ignore lower probabilities is explained in detail at the end of Sec. 2, but the basis is easy to state. The upper probabilities in the situations we consider are nonmonotonic

Reprinted from *Foundation of Physics*, **21** (1991), 1479–1499.
It is a pleasure to dedicate this paper to Karl Popper in celebration of his 90th birthday. The first author has known Popper for more than three decades, and has profited much from their discussion of many different topics, among which have been the foundations of probability and the foundations of quantum mechanics, both central to the present paper.

and for such upper probabilities the standard definition of lower probabilities does not work.

From a more general philosophical standpoint it is clear that views about hidden variables are very much conditioned on having a complete probabilistic account of the phenomena considered. Almost certainly, most physicists and philosophers will not find the existence of upper probabilities, without corresponding probabilities, physically appealing. Certainly as we ourselves have pointed out in earlier articles on upper and lower probabilities (Suppes and Zanotti, 1977, 1989), the natural questions about conditional upper and lower probabilities and independence are much more complicated than in the standard probabilistic framework. On the other hand, there has been very extensive consideration of weakening classical logic to various special forms of quantum logic. There may be for the Bell-type situations equally good arguments for keeping the logic standard, but weakening the probability requirements in the direction we develop in this paper.

To illustrate the development of ideas, in the next section we consider the standard elementary example of three random variables that do not have a joint probability distribution, namely three random variables with values ± 1 and with the correlation of any pair being -1. We construct an upper measure for these three random variables and also show the existence of a hidden variable – what we call a *generalized* random variable – with respect to the upper measure. In the following section we prove a theorem about the existence of generalized common causes, i.e., hidden variables, for finite sequences of pairwise correlated random variables. In Sec. 3 we apply the same techniques to a standard example of quantum mechanics that does not satisfy the Bell inequalities. We also generalize this quantum mechanical example to a theorem about such cases.

1. THREE RANDOM VARIABLES WITH MAXIMUM NEGATIVE CORRELATIONS

As already indicated, we consider in this section three random variables X_1, X_2, X_3 with values ± 1 and expectations

$$E(X_1) = E(X_2) = E(X_3) = 0$$
$$\mathrm{Cov}(X_i, X_j) = -1, \qquad i \neq j.$$

We can express the covariance and correlation just in terms of the expectation because the standard deviations are one and the expectations are zero:

$$E(X_i X_j) = -1, \qquad i \neq j.$$

129

We use the notation

$$p_{i\bar{j}} = P(X_i = 1, X_j = -1), \text{etc.}$$

So

$$p_{i\bar{j}} = p_{\bar{i}j} = \tfrac{1}{2}, \quad i \neq j$$
$$p_{ij} = p_{\bar{i}\bar{j}} = 0.$$

This implies, to fit the correlations,

$$p_{i\bar{j}}^* = \tfrac{1}{2}, \qquad p_{\bar{i}j}^* = \tfrac{1}{2}$$
$$p_{ij}^* = 0, \qquad p_{\bar{i}\bar{j}}^* = 0.$$

In the previous four equations we have used standard notation for the upper probabilities, a superscript star. More generally, we have the following axioms on upper probability, embodied in a definition.

DEFINITION 1 *Let Ω be a nonempty set, $\mathcal{F}$ a Boolean algebra on Ω, and P^* a real-valued function on $\mathcal{F}$. Then $\Omega = (\Omega, \mathcal{F}, P^*)$ is an* upper probability space *if and only if for every A and B in $\mathcal{F}$*.

1. $0 \leqslant P^*(A) \leqslant 1$;
2. $P^*(\emptyset) = 0$ and $P^*(\Omega) = 1$;
3. If $A \cap B = \emptyset$, then $P^*(A \cup B) \leqslant P^*(A) + P^*(B)$.

Axiom 3 on finite subadditivity could be strengthened to σ-subadditivity, but we are not concerned with that issue here.

To be perfectly clear about our notation, note that

$$p_{i\bar{j}}^* = P^*(X_i = 1, X_j = -1).$$

Since "mixed" $i\bar{j}$ or $\bar{i}j$ never occur in p_{123}^* or $p_{\bar{1}\bar{2}\bar{3}}^*$, we may set

$$p_{123}^* = p_{\bar{1}\bar{2}\bar{3}}^* = 0.$$

By symmetry and to satisfy subadditivity – e.g., $p_{1\bar{2}}^* \leqslant p_{1\bar{2}3}^* + p_{1\bar{2}\bar{3}}^*$, since

$$p_{i\bar{j}}^* = p_{\bar{i}j}^* = \tfrac{1}{2}, \qquad \text{for} \quad i \neq j$$

we set the remaining 6 triples at $\tfrac{1}{4}$:

(1) $$p_{12\bar{3}}^* = p_{1\bar{2}3}^* = p_{\bar{1}23}^* = p_{1\bar{2}\bar{3}}^* = p_{\bar{1}2\bar{3}}^* = p_{\bar{1}\bar{2}3}^* = \tfrac{1}{4}.$$

Notice that P^* is nonmonotonic for $p_{1\bar{2}3}^* > p_{12} = 0$. We examine this phenomenon in detail in Sec. 2.

130

We now define in the expected manner the upper expectation of a random variable which we express here for X_1:

$$(2) \qquad E^*(X_1) = \sum xp^*(x)$$
$$= 1(p^*_{123} + p^*_{1\bar{2}3} + p^*_{12\bar{3}}) + (-1)(p^*_{\bar{1}23} + p^*_{1\bar{2}\bar{3}} + p^*_{\bar{1}2\bar{3}})$$
$$= 0.$$

By symmetry

$$(3) \qquad E^*(X_i) = E(X_i) = 0, \qquad \text{for} \quad i = 1, 2, 3.$$

It is also obvious how we define the upper expectation of the product of two random variables:

$$E^*(X_i X_j) = \sum x_i x_j p^*(x_i x_j), \qquad i \neq j$$
$$= (-1)p^*_{i\bar{j}} + (-1)p^*_{\bar{i}j}$$
$$= -\tfrac{1}{2} + -\tfrac{1}{2}$$
$$= -1.$$

So the correlations are preserved:

$$(4) \qquad E^*(X_i X_j) = E(X_i X_j) = -1, \qquad i \neq j.$$

Hidden variable. We now turn to the existence of a hidden variable for three random variables. Given

$$(5) \qquad E(X_i X_j) = -1, \qquad 1 \leqslant i < j \leqslant 3$$

we want to find a hidden variable λ so that

$$(6) \qquad E^*(X_i X_j \mid \lambda = \lambda) = E^*(X_i \mid \lambda = \lambda)E^*(X_j \mid \lambda = \lambda)$$

where the upper conditional expectation for any random variable X is defined in the obvious conditional way:

$$(7) \qquad E^*(X \mid \lambda = \lambda) = \sum x P^*(X = x \mid \lambda = \lambda)$$

with, of course, upper conditional probability having the usual definition

$$(8) \qquad P^*(X = x \mid \lambda = \lambda) = P^*(X = x, \lambda = \lambda)/P^*(\lambda = \lambda)$$

for all values λ.

Moreover, we also want to satisfy

$$(9) \qquad E(X_i X_j \mid \lambda = \lambda) = E^*(X_i X_j \mid \lambda = \lambda).$$

There is no uniqueness requirement on the hidden variable λ. Many different characterizations will work. The most transparent construction in our judgment

131

is the deterministic one that mirrors as closely as possible the simultaneous possible values of the three random variables X_1, X_2, and X_3. So we choose

$$\lambda = (\pm 1, \pm 1, \pm 1)$$

where the ith coordinate of the vector corresponds to the values of X_i. When we consider the correlation of X_1 and X_2, for instance, we write $(1, 1, \cdot) = \{(1, 1, 1), (1, 1, -1)\}$.

Then

$$(10) \qquad p_{1\bar{2}}^* = P^*(X_1 = 1, X_2 = -1, \lambda \in (1, -1, \cdot))$$

since λ is deterministic.

Since $p_{1\bar{2}} = p_{1\bar{2}}^*$, we have

$$(11) \qquad P(X_1 = 1, X_2 = -1 \mid \lambda \in (1, -1, \cdot)) = 1$$

and

$$P(\lambda \in (1, -1, \cdot)) = P^*(\lambda \in (1, -1, \cdot)) = \tfrac{1}{2}$$

but, of course, we must have

$$P^*(\lambda = (1, -1, 1)) = P^*(\lambda = (1, -1, -1)) = \tfrac{1}{4},$$

for the upper probabilities of the values of λ mirror the values of X_1, X_2, X_3, taken together.

We now have the same strategy as in (1), for assigning upper probabilities of $\tfrac{1}{4}$ to each pair generated by terms like that of (9). Thus

$$p^*(1, \bar{2}, (1, -1, \cdot)) = \tfrac{1}{2} \leqslant p^*(1, \bar{2}, 3, (1, -1, 1))$$
$$+ p^*(1, \bar{2}, \bar{3}, (1, -1, -1)) = \tfrac{1}{4} + \tfrac{1}{4}.$$

The conditional expectations for the pairs $X_i X_j$, $i \neq j$ then all have the expected deterministic values, for

$$E^*(X_i X_j \mid \lambda = \lambda) = E(X_i X_j \mid \lambda = \lambda).$$

More important is the factorization. For example,

$$E(X_1 X_2 \mid \lambda \in (1, -1, \cdot)) = E(X_1 \mid \lambda \in (1, -1, \cdot)) E(X_2 \mid \lambda \in (1, -1, \cdot))$$
$$= 1 \cdot (-1)$$
$$= -1.$$

(Notice that we have written E, since here $E^* = E$ in the restricted cases considered.) Because the hidden variable λ is deterministic, all aspects of the upper probabilities are entirely reflected in the upper probabilities for the values of λ. Thus

$$p^*(\pm 1, \pm 1, \pm 1) = \tfrac{1}{4} \text{ except } p^*(1, 1, 1) = p^*(-1, -1, -1) = 0$$

$$p^*(1, -1, \cdot) = \tfrac{1}{2}, \text{ etc.}$$

$$p^*(1, \cdot, \cdot) = \tfrac{1}{2}, \text{ etc.}$$

The hidden variable constructed above mirrors quite directly the upper probability measure on the space of possible outcomes of the three random variables X_1, X_2, and X_3. As can be seen from the construction, if we only consider a pair, say X_1 and X_2 with correlation -1, then a hidden variable that is random variable can be constructed and no generalization to an upper measure is needed. Additional requirements of symmetry with respect to λ can change the situation, as we now show.

Symmetry with respect to λ. In an earlier paper, we proved the following theorem that gives a necessary and sufficient condition for two-valued exchangeable random variables to have "identical" causes in the sense of conditional expectation.

THEOREM (Suppes and Zanotti, 1980) *Let X and Y be two-valued random variables, for definiteness, with possible values 1 and -1, and with positive variances, i.e., $\sigma(X), \sigma(Y) > 0$. In addition, let X and Y be exchangeable, i.e.,*

$$P(X = 1, Y = -1) = P(X = -1, Y = 1).$$

Then a necessary and sufficient condition that there exist a hidden variable λ such that $E(XY \mid \lambda = \lambda) = E(X \mid \lambda = \lambda)E(Y \mid \lambda = \lambda)$ and $E(X \mid \lambda = \lambda) = E(Y \mid \lambda = \lambda)$ for every value λ (except possibly on a set of measure zero) is that the correlation of X and Y be nonnegative.

Random variables X_1 and X_2 as defined earlier with correlation -1 satisfy the hypothesis of this theorem, and so they can have no hidden variable within standard probability theory satisfying the symmetry condition of causality:

$$(12) \qquad E(X_1 \mid \lambda = \lambda) = E(X_2 \mid \lambda = \lambda).$$

On the other hand, we can find a hidden variable with an upper probability measure satisfying (12), as well as the standard factorization, but now in terms of *upper* conditional expectations, which we defined earlier.

The specification of p^* is as follows, where the subscript 3 refers to values of λ and, as before, 3 indicates the value $\lambda = 1$ and $\bar{3}$ the value $\lambda = -1$, with only two values of λ being needed for this analysis:

$$p^*_{ijk} = \tfrac{1}{4}$$

for all 8 combinations of ijk.

133

Also

$$p_{1..} = p_{.2.} = p_{..3} = \tfrac{1}{2}$$
$$p_{12.} = p_{\bar{1}2.} = 0$$
$$p_{1\bar{2}.} = p_{\bar{1}\bar{2}.} = \tfrac{1}{2}.$$

The rest of the specifications to satisfy the constraints of the situation are obvious, although λ is a stochastic rather than deterministic hidden variable, so it is easy to compute that

$$E^*(X_1 \mid \lambda = \pm 1) = E^*(X_2 \mid \lambda = \pm 1) = 0$$
$$E^*(X_1 X_2 \mid \lambda = \pm 1) = 0,$$

which satisfy the condition required for symmetry in "the" cause of a negative correlation. But as can be seen from the value of p_{ijk}^*, we must construct p^* to be nonmonotonic.

2. GENERALIZED RANDOM VARIABLES AS GENERALIZED COMMON CAUSES

The hidden variable λ of the previous section suggests at once a generalization of the standard notion of a random variable. As the example suggests, it will be convenient for a generalized random variable to take as values vectors of real numbers of a given dimension, rather than simply real numbers. But this is not the real point of our use of the term *generalized*; it is rather the generalization from a random variable's having a probability to having an upper probability.

Let $\Omega = (\Omega, \mathcal{F}, P^*)$ be an upper probability space and let λ be a function from Ω to Re^k such that for every vector $(b_1, \ldots, b_k)$ the set

$$\{\omega : \omega \in \Omega \,\&\, \lambda_i(\omega) \leqslant b_i, i = 1, \ldots, k\}$$

is in $\mathcal{F}$. Then λ is a *generalized random variable (with respect to Ω)*.

Extending the earlier example, we can then prove that for any finite set of two-valued correlated random variables, there exists an underlying generalized random variable that makes each of the correlated pairs conditionally independent. Put in more general philosophical language, we prove the existence of a generalized common cause.

This result extends our earlier result (Suppes and Zanotti, 1981) that a joint probability of all the random variables is necessary and sufficient for the existence of a common cause. Of course, the extension is at the expense of using the weaker concept of an upper probability measure.

Before stating the theorem, we introduce some concepts and notation that are needed. Let Ω be the space on which the random variables $X_1, \ldots, X_n$

are defined and let $\mathcal{F}$ be the given algebra of events, i.e., subsets of Ω. We require that $\mathcal{F}$ contain as a subalgebra $\mathcal{F}^*$, the algebra of cylinder sets of Ω defined by the sequence of values of the random variables $X_1, \ldots, X_n$. First, we characterize any index set (of dimension no greater than n) by two disjoint sets of positive integers I and J, with the largest integer in either set being no greater than n. The set I is the set of indices i for which $X_i = 1$ and the set J is the set of indices i for which $X_i = -1$. So a pair (I, J) uniquely defines an index set, an element of $\mathcal{F}^*$, relative to $X_1, \ldots, X_n$.

Note that if $I \cup J = \emptyset$, then $(I, J) = \Omega$, and if $\mid I \cup J \mid = n$, i.e., the cardinality of the set $I \cup J$ is n, then (I, J) is an atom of $\mathcal{F}^*$. To illustrate the notation concretely, if $n = 3$, then $(\{1\}, \{3\}) = (1, \cdot, -1)$, in our earlier notation.

It is easy to show explicitly that the algebra $\mathcal{F}^*$ of cylinder sets results from taking the closure under union and complementation of the index sets. So if (I, J) is an index, then its complement is a cylinder set, and if (I_1, J_1) and (I_2, J_2) are two index sets, then $(I_1, J_1) \cup (I_2, J_2)$ is a cylinder set.

What is central to the theorem is to begin by assuming only pairwise probability functions of the random variables and then to construct an upper measure P^* on $\mathcal{F}^*$. The measure constructed is in general not unique.

THEOREM 1 [Generalized Common Causes] *Let $X_1, \ldots, X_n$ be two-valued* (± 1) *random variables whose common domain is a space Ω with an algebra $\mathcal{F}$ of events that includes the subalgebra $\mathcal{F}^*$ of cylinder sets of dimension n defined above. Also, let pairwise probability functions $P_{ij}, 1 \leqslant i < j \leqslant n$, compatible with the single functions $P_i, 1 \leqslant i \leqslant n$, be given. Then there exists an upper probability space $\Omega = (\Omega, \mathcal{F}^*, P^*)$ and a generalized random variable λ on Ω to the set of n-dimensional vectors whose components are ± 1 such that for $1 \leqslant i < j \leqslant n$ and every value λ of λ.*

(i) $P^*(X_i = \pm 1, X_j = \pm 1) = P_{ij}(X_i = \pm 1, X_j = \pm 1)$;
(ii) $P^*(X_1 = \lambda_1, \ldots, X_n = \lambda_n) = P^*(\lambda_1 = \lambda_1, \ldots, \lambda_n = \lambda_n)$;
(iii) *λ is deterministic, i.e.,*

$$P(X_i = 1 \mid \lambda_i = 1) = 1$$

and

$$P(X_i = -1 \mid \lambda_i = -1) = 1;$$

(iv) $E(X_i X_j \mid \lambda = \lambda) = E(X_i \mid \lambda = \lambda) E(X_i \mid \lambda = \lambda)$.

Proof. We first set $P^*(\Omega) = 1$. As shown earlier, a pair (I, J) of mutually exclusive sets of positive integers uniquely defines an index set. In this notation,

135

by the hypothesis of the theorem we are given for a singleton or pair set – here the cardinality of $I \cup J$ is 1 or 2 – the probability, so we set

$$(13) \qquad P^*((I, J)) = P_{ij}((I, J)), \qquad 1 \leqslant |I \cup J| \leqslant 2.$$

Consider now any index set (I, J) such that $|I \cup J| > 2$. We define recursively

$$(14) \qquad P^*((I, J)) = \tfrac{1}{2} \max_{I', J'} (P^*(I', J), P^*(I, J'))$$

where, if $I \neq \emptyset$, then $I' \subseteq I, |I'| = |I| - 1$; if $J \neq \emptyset$, then $J' \subseteq J$, $|J'| = |J| - 1$; if $I = \emptyset$, then $I' = \emptyset$; and if $J = \emptyset$, then $J' = \emptyset$.

We now prove that P^* is subadditive on the index sets. We assert that for any $i \notin I \cup J$ with $|I \cup J| < n$

$$(15) \qquad P^*((I, J)) \leqslant P^*((I \cup \{i\}, J)) + P^*((I, J \cup \{i\})).$$

(Of course if $|I \cup J| = n$, then (I, J) is an atom.)

From (14), we infer, for $i \notin I \cup J$

$$P^*((I \cup \{i\}, J)) = \tfrac{1}{2} \max P^*((I \cup \{i\})', J), P^*((I \cup \{i\}, J'))$$

where the prime operation and max are as defined for (14). Thus

$$(16) \qquad P^*(I \cup \{i\}, J) \geqslant \tfrac{1}{2} P^*((I, J))$$

and by a similar argument

$$(17) \qquad P^*((I, J \cup \{i\})) \geqslant \tfrac{1}{2} P^*((I, J)).$$

Inequality (15) follows at once from (16) and (17), so P^* is an upper measure on the index sets of $(\Omega, \mathcal{F}^*)$. We now extend P^* to all of the cylinder sets (of dimension not greater than n), i.e. to all of $(\Omega, \mathcal{F}^*)$. To avoid heavy notation we first describe informally this extension. For any cylinder set A that is not an index set, we consider all partitions of the set in terms of index sets. For each partition we take the arithmetic sum of the upper measure P^* of the sets in the partition. We now take the min of P^* over these partitions as $P^*(A)$. In symbols

$$(18) \qquad P^*(A) = \min_{\Pi(A)} \sum_{L_i \in \Pi(A)} P^*(L_i)$$

where $\Pi(A)$ is a partition of A in terms of index sets L_i. The min operation guarantees that the axiom of subadditivity will be satisfied by A, i.e., we have at once from (18)

$$(19) \qquad P^*(A) \leqslant \sum_{L_i \in \Pi(A)} P^*(L_i).$$

136

The same argument applies to Ω, so that

$$P^*(\Omega) = P^*(A \cup \bar{A}) \leqslant P^*(A) + P^*(\bar{A})$$

and from (18) it is also clear that

$$0 \leqslant P^*(A) \leqslant 1$$

which completes the proof that P^* is an upper probability on $(\Omega, \mathcal{F}^*)$. Equation (13) guarantees satisfaction of (i) of the theorem.

The hidden variable λ, which is a function from Ω to the set of n-dimensional vectors whose components are ± 1, is defined to be deterministic, i.e., for $1 \leqslant i \leqslant n$:

$$P(X_i = 1 \mid \lambda_i = 1) = 1$$

and

$$P(X_i = -1 \mid \lambda_i = -1) = 1$$

so the upper probability function for λ is the same as that of $X_1, \ldots, X_n$, which establishes (ii) and (iii).

Finally, since λ is deterministic, the conditional independence of X_i and X_j, $i \neq j$, given $\lambda = \lambda$, follows at once, which establishes (iv) and completes the proof of the theorem.

Status of monotonicity. A familiar strong axiom for upper measures in the context of statistical inference is the monotonicity axiom:

$$\text{If } A \subseteq B, \text{ then } P^*(A) \leqslant P^*(B).$$

This axiom is fundamental for Walley (1991), Dempster, Shafer, and others. Unfortunately, it is violated already by the example given in the preceding section of three random variables having pairwise correlations of -1. When we say "violated," we mean it is easy to show that no upper measure compatible with the given pairwise probability distributions can satisfy the axiom of monotonicity. Intuitively speaking, satisfaction of this axiom in the case of three two-valued variables implies there is a probability measure dominated by the upper measure and compatible with the pairwise distributions, so that an upper measure that is monotonic cannot be constructed for cases where no probability distribution exists. The following two theorems summarize these facts.

THEOREM 2 [Monotonicity Implies Probability] *Let X_1, X_2, and X_3 be two-valued (± 1) random variables with $E(X_i) = 0$, $i = 1, 2, 3$, such that there is a monotonic upper probability function compatible with the given correlations $E(X_i X_j)$, $1 \leqslant i < j \leqslant 3$. Then there exists a joint probability function of X_1, X_2, and X_3.*

Proof. A necessary and sufficient condition for the existence of a joint probability distribution of three random variables satisfying the hypothesis of the theorem is given in Suppes and Zanotti (1981):

(20)
$$-1 \leqslant E(X_1X_2) + E(X_2X_3) + E(X_1X_3)$$
$$\leqslant 1 + 2\min(E(X_1X_2), E(X_2X_3), E(X_1X_3)).$$

For brevity of notation we write $\rho_{ij} = E(X_iX_j)$, and we prove that when P^* is monotonic, (20) follows.

Since $E(X_i) = 0$, it is obvious that

$$p_{ij} - p_{i\bar{j}} = \frac{\rho_{ij}}{2}$$
$$p_{ij} + p_{i\bar{j}} = \tfrac{1}{2}$$

so

(21)
$$4p_{ij} = 1 + \rho_{ij}$$
$$4p_{i\bar{j}} = 1 - \rho_{ij}.$$

By subadditivity we then have

(22)
$$4p_{1\bar{2}.} = 1 - \rho_{12} \leqslant 4p^*_{12\bar{3}} + 4p^*_{1\bar{2}\bar{3}}.$$

And by monotonicity

(23)
$$4p^*_{1\bar{2}3} \leqslant 4p_{1.3} = 1 + \rho_{13}$$
$$4p^*_{12\bar{3}} \leqslant 4p_{.2\bar{3}} = 1 + \rho_{23}.$$

From (22) and (23)

$$1 - \rho_{12} \leqslant 1 + \rho_{23} + 1 + \rho_{13}$$

and so

$$-1 \leqslant \rho_{12} + \rho_{23} + \rho_{13}.$$

Second, without loss of generality, we may assume

(24)
$$\rho_{12} = \min(\rho_{12}, \rho_{23}, \rho_{13}).$$

But using (21) again and monotonicity

$$1 + \rho_{13} \leqslant 4p^*_{123} + 4p^*_{1\bar{2}3}$$
$$\leqslant 1 + \rho_{12} + 1 - \rho_{23}$$

so

(25)
$$\rho_{23} + \rho_{13} \leqslant 1 + \rho_{12}$$

138

but (24) and (25) imply

$$\rho_{12} + \rho_{23} + \rho_{13} \leqslant 1 + 2 \min(\rho_{12}, \rho_{23}, \rho_{13})$$

which completes the proof.

The proof of the following theorem on monotonicity is an immediate consequence of the proof of Theorem 2.

THEOREM 3 [Nonmonotonicity] *Let $X_1, X_2,$ and X_3 be two-valued (± 1) random variables with $E(X_i) = 0, i = 1, 2, 3,$ such that there is no joint probability distribution compatible with the correlations $E(X_i X_j), 1 \leqslant i < j \leqslant 3$. Then any upper measure P^* compatible with the given correlations cannot satisfy the axiom of monotonicity.*

The consequences of nonmonotonicity reach even further. In the usual statistical applications, where it is assumed that P^* is monotonic, the lower probability of any event A is defined by

$$(26) \qquad\qquad P_*(A) = 1 - P^*(\bar{A})$$

and under the assumption of monotonicity it can be proved that P_* is superadditive, i.e., if $A \cap B = \emptyset$, then

$$P_*(A) + P_*(B) \leqslant P_*(A \cup B).$$

We now prove that P_* when defined by (26) cannot be superadditive if P^* is nonmonotonic, which means that the standard relationship (26) cannot work in the environments we are concerned with.

THEOREM 4 *Let $(\Omega, \mathcal{F}, P^*)$ be an upper probability space such that P^* is nonmonotonic. Then the lower probability P_* defined by (26) is not superadditive.*

Proof. By the hypothesis of the theorem there are events A and B such that $A \subseteq B$ but $P^*(B) < P^*(A)$. Clearly $A \neq B$ for otherwise we would have the immediate contradiction that $P^*(A) < P^*(A)$. So there is a $C \neq \emptyset$ such that $A \cap C = \emptyset$ and

$$A \cup C = B.$$

Suppose now that P_* is superadditive. Then since $\bar{A} = \bar{B} \cup C$

$$P_*(\bar{B}) + P_*(C) \leqslant P_*(\bar{A})$$

so

$$1 - P^*(B) + 1 - P^*(\bar{C}) \leqslant 1 - P^*(A)$$

139

and thus

$$1 + P^*(A) \leqslant P^*(B) + P^*(\bar{C})$$

but $P^*(\bar{C}) \leqslant 1$ by Axiom 1 of Definition 1 for upper probability measures, whence

$$P^*(A) \leqslant P^*(B)$$

contrary to our initial hypothesis, which proves the theorem.

3. HIDDEN VARIABLES IN QUANTUM MECHANICS

We use the standard notation familiar in the Bell inequalities which we review very briefly. For definiteness, but not required, we can think of a Bell-type experiment in which we are measuring spin for particle A and for particle B. More generally, we may think of A and B as being the location of measuring equipment and we observe individual particles or a flux of particles at each of the sites. Here we will think of individual particles because the analysis is simpler. The measuring apparatus is such that along the axis connecting A and B we have axial symmetry and consequently we can describe the position of the measuring apparatus just by the angle of the apparatus A or B in the plane perpendicular to the axis. We use the notation ω_A and ω_B for these angles. The basic form of the locality assumption is shown in terms of the following expectation:

(27) $$E(M_A \mid \omega_A, \omega_b, \lambda) = E(M_A \mid \omega_A, \lambda)$$

What this means is the expectation of the measurement M_A of spin of a particle in the apparatus in position A, given the two angles of measurement for apparatus A and B as well as λ, is equal to the expectation without any knowledge of the apparatus angle ω_B of B. This is a reasonable causal assumption and is a way of saying that what happens at B should have no direct causal influence on what happens at A. On the other hand, we have the following theoretical result for spin, well confirmed in principle for the case where the measuring apparatuses are both set at the same angle:

(28) $$P(M_A = -1 \mid \omega_A = \omega_B = \alpha \,\&\, M_B = 1) = 1$$

If the angles of the apparatus are set the same, we have a deterministic result in the sense that the observation of spin at B will be the opposite at A, and conversely. Here we are letting 1 correspond to spin $\frac{1}{2}$ and -1 correspond to spin $-\frac{1}{2}$. What Bell showed is that on the assumption there exists a hidden variable, four related inequalities can be derived for settings A and A' and B and B' for the measuring apparatus. We have reduced the notation here in the following way in writing the inequalities. First, instead of writing M_A we write simply A, and second, instead of writing $\text{Cov}(A, B)$ for covariance, which in this case will be the same as the correlation, of the measurement at A and the measurement at B, we write simply AB. With this understanding about the conventions of the notation, we then have as a consequence of the assumption of a hidden variable the following set of inequalities, which in the exact form given here are due to Clauser, Horne, Shimony, and Holt (1969):

(29)
$$-2 \leqslant AB + AB' + A'B - A'B' \leqslant 2$$
$$-2 \leqslant AB + AB' - A'B + A'B' \leqslant 2$$
$$-2 \leqslant AB - AB' + A'B + A'B' \leqslant 2$$
$$-2 \leqslant -AB + AB' + A'B + A'B' \leqslant 2.$$

Quantum mechanics does not satisfy these inequalities in general. To illustrate ideas, we take as a particular case the following:

$$AB - AB' + A'B + A'B' < -2.$$

We choose

$$AB = A'B' = -\cos 30° = -\frac{\sqrt{3}}{2}$$
$$AB' = -\cos 60° = -\frac{1}{2}$$
$$A'B = -\cos 0° = -1.$$

So

$$-\frac{\sqrt{3}}{2} + \frac{1}{2} - 1 - \frac{\sqrt{3}}{2} < -2$$

since from quantum mechanics $\text{Cov}(AB) = -\cos(\text{angle } AB)$.

As we can see from the example of Sec. 2, we need only consider an upper probability for values of λ.

Here

$$\lambda = \begin{pmatrix} A & A' & B & B' \\ \pm 1, & \pm 1, & \pm 1, & \pm 1 \end{pmatrix}.$$

First, we must compute the probabilities for the pairs with given correlations. So

$$p(1, \cdot, \cdot, \cdot) = p(-1, \cdot, \cdot, \cdot) = \tfrac{1}{2}$$

since $E(A) = 0$, etc.

Now

$$AB = -\frac{\sqrt{3}}{2}$$

141

so

$$-\frac{\sqrt{3}}{2} = p(1, \cdot, 1, \cdot) + p(-1, \cdot, -1, \cdot) - p(1, \cdot, -1, \cdot) - p(-1, \cdot, 1, \cdot).$$

But by symmetry

$$p(1, \cdot, 1, \cdot) = p(-1, \cdot, -1, \cdot)$$

and

$$p(1, \cdot, -1, \cdot) = p(-1, \cdot, 1, \cdot).$$

So solving, we obtain

$$4p(1, \cdot, 1, \cdot) - 1 = \frac{\sqrt{3}}{2}$$

and

$$p(1, \cdot, 1, \cdot) = -\frac{\sqrt{3}}{8} + \frac{1}{4}$$

$$p(1, \cdot, -1, \cdot) = \frac{\sqrt{3}}{8} + \frac{1}{4}.$$

Similarly for $A'B' = -\sqrt{3}/2$

$$p(\cdot, 1, \cdot, 1) = -\frac{\sqrt{3}}{8} + \frac{1}{4}$$

$$p(\cdot, 1, \cdot, -1) = \frac{\sqrt{3}}{8} + \frac{1}{4}.$$

Next, $AB' = -\frac{1}{2}$, so

$$4p(1, \cdot, \cdot, 1) - 1 = -\frac{1}{2}$$
$$p(1, \cdot, \cdot, 1) = \frac{1}{8}$$
$$p(1, \cdot, \cdot, -1) = \frac{3}{8}.$$

Since $A'B = -1$

$$4p(\cdot, 1, 1, \cdot) - 1 = -1$$
$$p(\cdot, 1, 1, \cdot) = 0$$
$$p(\cdot, 1, -1, \cdot) = \frac{1}{2}$$

We are now in a position to compute the triples. We restrict ourselves to $p^*_{i \cdot jk}$ to illustrate the method. So, for example, consider the triple $p^*_{1 \cdot 11} = P^*(A = 1, B = 1, B' = 1)$. We have at once

$$\tfrac{1}{8} = p^*_{1 \cdot \cdot 1} \leqslant p^*_{1 \cdot 11} + p^*_{1 \cdot -11}.$$

142

To get as close as we can to a probability, we take

$$p^*_{1\cdot11} = 0 \qquad \text{and} \qquad p^*_{1\cdot-11} = \tfrac{1}{8}$$

and by symmetry

$$p^*_{-1\cdot1-1} = \tfrac{1}{8} \qquad \text{and} \qquad p^*_{-1\cdot-1-1} = 0.$$

By similar arguments, we set

$$p^*_{1\cdot1-1} = p^*_{-1\cdot-11} = \frac{1}{4} - \frac{\sqrt{3}}{8}$$

$$p^*_{-1\cdot11} = p^*_{1\cdot-1-1} = \frac{1}{8} + \frac{\sqrt{3}}{8}.$$

The remaining upper probabilities for this example may be found by similar lines of argument.

We prove a theorem about quantum mechanical covariances that follows directly from the theorem on generalized common causes.

THEOREM 5 [Existence of Hidden Variables] *Let* $AB, AB', A'B,$ *and* $A'B'$ *be any four quantum mechanical covariances, which will in general not satisfy the Bell inequalities. Then there is an upper probability* P^* *consistent with the given covariances and a generalized hidden variable* λ *with* P^* *such that, for every value* λ *of* $\boldsymbol{\lambda}$,

$$E(AB \mid \boldsymbol{\lambda} = \lambda) = E(A \mid \boldsymbol{\lambda} = \lambda)E(B \mid \boldsymbol{\lambda} = \lambda)$$

and similarly for $AB', A'B,$ *and* $A'B'$.

Proof. Take the unspecified correlations AA' and BB' to have any values between -1 and 1. Then apply Theorem 1, with $n = 4$.

Corresponding to Theorem 2 showing that monotonicity of P^* implies existence of a probability distribution, we may derive the Bell inequalities just from existence of a monotonic upper probability. We believe this derivation is new in the literature. The similarity to Theorem 2 is evident, since satisfaction of the Bell inequalities (29) implies existence of a joint probability distribution of $A, A', B,$ and B'.

THEOREM 6 [Monotonicity Implies Bell Inequalities] *Let* $A, A', B,$ *and* B' *be two-valued* (± 1) *random variables with expectation* $E(A) = E(A') = E(B) = E(B') = 0$ *such that there is a monotonic upper probability function compatible with the given correlations* $AB, AB', A'B,$ *and* $A'B'$. *Then the given covariances satisfy the Bell inequalities.*

Proof. Using the relations established for the proof of Theorem 2 when $E(X_i) = 0$, we have first by subadditivity

(30) $$1 - AB \leqslant 4p(11 - 1\cdot) + 4(1 - 1 - 1\cdot).$$

Applying subadditivity again, we get

(31) $$4p(11 - 1\cdot) \leqslant 4p(11 - 11) + 4p(11 - 1 - 1).$$

By monotonicity, we have at once

$$4p(1 - 1 - 1\cdot) \leqslant 4p(\cdot - 1 - 1\cdot) = 1 + A'B$$
(32) $$4p(11 - 11) \leqslant 4p(1 \cdot \cdot 1) = 1 + AB'$$
$$4p(11 - 1 - 1) \leqslant 4p(\cdot 1 \cdot -1) = 1 - A'B'.$$

Combining (30), (31), and (32), we obtain

$$1 - AB \leqslant 1 + AB' + 1 + A'B + 1 - A'B'$$

which is equivalent to

(33) $$-2 \leqslant AB + AB' + A'B - A'B'.$$

Secondly, by subadditivity

(34) $$4p(\cdot 1 \cdot -1) = 1 - A'B' \leqslant 4p^*(11 \cdot -1) + 4p^*(-11 \cdot -1).$$

And again by subadditivity

(35) $$4p^*(-11 \cdot -1) \leqslant 4p^*(-111 - 1) + 4p^*(-11 - 1 - 1).$$

But by monotonicity

$$4p^*(11 \cdot -1) \leqslant 4p(1 \cdot \cdot - 1) = 1 - AB'$$
(36) $$4p^*(-111 - 1) \leqslant 4p(-1 \cdot 1\cdot) = 1 - AB$$
$$4p^*(-11 - 1 - 1) \leqslant 4p(\cdot 1 - 1\cdot) = 1 - A'B.$$

Combining (34), (35), and (36), we have

$$1 - A'B' \leqslant 1 - AB + 1 - AB' + 1 - A'B$$

which is equivalent to

(37) $$AB + AB' + A'B - A'B' \leqslant 2.$$

Inequalities (33) and (37) are the first line of the Bell inequalities (29) given earlier. By completely similar arguments we may derive the other three.

The proof of Theorem 6 also has as an immediate consequence a negative theorem about monotonicity, similar to Theorem 3 for three random variables.

THEOREM 7 [Nonmonotonicity] *Let $AB, AB', A'B$, and $A'B'$ be quantum mechanical covariances that do not satisfy Bell's inequalities. Then any upper measure P^* compatible with the given correlations cannot be monotonic.*

144

4. DATA TABLES AND NONMONOTONIC UPPER PROBABILITIES

Not only in quantum mechanics but in standard statistical analysis the impossibility of the data not having a joint probability distribution implies at once a severe restriction on the kind of experiments that can be performed.

Consider first the classical general case discussed in Sec. 1 of three two-valued random variables whose pairwise correlations algebraically sum to less than -1. We can infer immediately that any experiments yielding such results could not have been ones where X_1, X_2, and X_3 were simultaneously observed. If they had, the joint relative frequencies arising from the data table of the experiment would necessarily be consistent with a joint probability distribution. Put more strongly, the relative frequencies of the 8 types of triples $(\pm 1, \pm 1, \pm 1)$ that could be observed in such an experiment constitute themselves a probability distribution.

The same remarks apply to the quantum mechanical experiments of the Bell type. Here the problem is even more troublesome, because the quantum mechanical covariances that violate the Bell inequalities and that are consistent only with a nonmonotonic upper probability are confirmed both experimentally and theoretically. So it becomes a quantum mechanical theoretical constraint that no experiment with individual particles could be performed to observe simultaneously the values of A, A', B, and B'.

Because such simultaneous observation is impossible, we are not able to use ordinary probabilistic analysis to understand more thoroughly the phenomena being studied. For example, one of the most fundamental steps of analysis in probability is that of conditionalizing. In ordinary terms, having measured a correlation AB, and having also observed the correlation AB' we could naturally go on to study such conditional expectations as

$$E^*(AB \mid B' = 1)$$

which we express here in terms of the upper expectation. But, of course, this step cannot be taken even though we have available in theory this kind of conditional expectation as defined in Eq. (7). The step cannot be taken, for no simultaneous individual observation data of A, B, and B' can be collected, even in principle, consistent with quantum mechanics – we mean, of course, triples of observations of the form $(\pm 1, \pm 1, \pm 1)$.

The hidden variables we have constructed with nonmonotonic upper probabilities are indeed hidden. Whether or not anything further can be done with them in the context of quantum phenomena remains to be seen. Skepticism is warranted, for any direct application would go beyond standard quantum mechanics.

145

REFERENCES

Clauser, J. F., Horne, M. A., Shimony, A., and Holt, R. A.: 1969, Proposed experiment to test local hidden-variable theories. *Phys. Rev. Lett*, **23**, 880–884.

Suppes, P., and Zanotti, M.: 1977, On using random relations to generate upper and lower probabilities. *Synthese*, **36**, 427–440 [chap. 3, this volume].

Suppes, P., and Zanotti, M.: 1980, A new proof of the impossibility of hidden variables using the principles of exchangeability and identity of conditional distribution. In P. Suppes (Ed.), *Studies in the Foundations of Quantum Mechanics*. East Lansing, Mich.: Philosophy of Science Association. pp. 173–191 [chap. 8, this volume].

Suppes, P., and Zanotti, M.: 1981, When are probabilistic explanations possible? *Synthese*, **48**, 191–199 [chap. 9, this volume].

Suppes, P., and Zanotti, M.: 1989, Conditions on upper and lower probabilities to imply probabilities. *Erkenntnis*, **31**, 323–345 [chap. 4, this volume].

Walley, P.: 1991, *Statistical Reasoning with Imprecise Probabilities*. London: Chapman.

III

Applications in education

13

Mastery learning of elementary mathematics: Theory and data

1. INTRODUCTION

In this final paper, which is the only one in this volume that has not been previously published, we develop in some detail a portion of the extensive work we have done together over nearly twenty years on the application of probabilistic models and methods to problems of curriculum and learning in a computer-based setting. It might seem that there is little connection between this work and what has been described in earlier parts of this volume. In our own mind, however, there is an intimate connection and considerable continuity. The general reason is easy to state. Although much of our work has been motivated by Bayesian considerations and the admiration we have for the pioneering work of Bruno de Finetti in developing concepts of subjective probability, we also recognize how complicated real applications of Bayesian ideas are. Even more generally, we recognize how complicated real applications of normative ideas are. The normative considerations that guide the work described here concern how to optimize the course of learning for each individual student. We pick as our example elementary mathematics, partly because there has been such a long history of research on the learning of elementary mathematics, especially arithmetic, throughout this century, beginning at least with the early work of Edward Thorndike, if not even earlier.

As one reflects on the problem of organizing a curriculum for optimizing individual learning, it is clear that general Bayesian considerations play only a small part. The real problem is to understand as thoroughly as possible the nature of student learning, and, second, to have detailed ideas about what is important for students to learn. The way in which Bayesian philosophy fits into this program is, on the other hand, clear. It is characteristic of a Bayesian viewpoint on real-world problems to be skeptical that final solutions of a simple kind, or of a kind that can be built on a bedrock of certainty, are seldom if ever to be found. In this same sense, any claim to have settled all the major questions

149

of student learning or of curriculum organization can only be greeted with proper Bayesian skepticism. What we can hope to do with these two facets of constructing a good computer-based course is to make some headway on understanding better than we do now how they should be organized and on what basis.

We also want to make clear that we do not think, even within a framework of subjective probability, it is possible to reach a definitive normative decision on curriculum organization. Conflict is inevitable and the methods of resolving the conflicts take us beyond Bayesian ideas to game-theoretic ones. In other words, concepts of bargaining, negotiation, and relative positions of power inevitably determine, even if implicitly, how problems of curriculum, especially problems of curriculum emphasis, will be resolved. The normative aspect of the work we have done is conditional in character. Given broad decisions about the nature of the curriculum coverage and the relative distributional emphasis of concepts and skills in that curriculum, we can then apply the methodology developed in this paper to optimize individual student learning. But – and it is important to stress this point – we do not present our work as a straightforward problem of optimization. The many problems of detailed analysis yet to be resolved do not make a direct optimization approach feasible. Much of what we do in this article is the application of specific models of learning that we can then compare with empirical data on student performance. The most elaborate formulation we have yet reached on the interlocking of curriculum organization and mastery criteria is to be found in the final section. Even this rather elaborate model-theoretic formulation is still much too simple in many respects.

In this article we describe extensive work at Stanford University and Computer Curriculum Corporation (CCC) over a number of years on computer-assisted instruction in elementary mathematics. Much of what we report here is relevant to other courses, for example, computer-assisted instruction in reading.

The article is organized along the following lines. Section 2 provides a brief review of the main components of mastery learning we have used and analyzed in our work. Section 3 is a detailed treatment of the theory of the learning models we use. Section 4 presents extensive data on how well the models fit some of the main probabilistic features of the students' responses to exercises in elementary mathematics.

Section 5 turns to the theory of global trajectories of students in a computer-based course. Section 6 then exhibits, mainly in the form of figures and graphs, extensive data on such trajectories in CCC courses on elementary mathematics and reading. Section 7 briefly describes the extension of the work on trajectories to prediction and intervention aimed at helping individual students meet agreed upon achievement goals.

Finally, Section 8 describes our new approach to mastery learning, which is being used in a computer-based course in elementary mathematics, Grades 1–8, at Stanford. This new course, unlike the CCC one, is not supplementary, but is a complete course of instruction aimed especially at students with above-average aptitude for learning mathematics. As a consequence the material on geometry goes far beyond what is ordinarily to be found in a standard elementary school mathematics curriculum.

2. COMPONENTS OF MASTERY LEARNING

There is a wide use of the concept of mastery learning at a qualitative or informal level in American schools. There are, of course, many different ways of conceiving a model or theory of mastery learning. We have found it natural to analyze our theoretical ideas in terms of six main components, each of which will be considered, although some in greater depth than others. The six components are: (1) curriculum distribution and dynamic ordering of concepts, (2) student distribution, (3) initial grade placement, (4) learning models for judging mastery, (5) forgetting models for assigning review, (6) decisions on tutorial intervention. We describe briefly each of these six components. All but (3) are examined in more detail in later sections. Our current conception of an overall model governing a student's movement in a curriculum is set forth in the final section. We sketch in the next few paragraphs the approach taken in the past at CCC. The following sections present detailed theoretical and empirical analysis of the various model components used. Here is the sketch of the six components.

Curriculum distribution. Not all concepts are equal in importance, so the expected time devoted to mastery should vary. Addition of positive integers, for example, is much less important in the fourth grade than addition of fractions. But how should the expected time be allocated and on what intellectual basis? There is, unfortunately, only a very limited literature on this important matter in the theory of curriculum. At the present time the most feasible approach is to use as initial data the curriculum guidelines set by various state and local school systems, and then to count the empirical frequency of exercises in various widely used textbooks. After pooling and smoothing these data, the next step is to use latency data to convert back to a distribution of exercise types organized by concept. This approach was described many years ago in Suppes (1967, 1972).

The current version of the elementary mathematics course at CCC, Math Concepts and Skills (MCS), is based on the strands that are shown in Table 1. The curriculum distribution for each half-grade level is shown in Table 2. The

Table 1. *The strands in Math Concepts and Skills (MCS).*

Strand name	Strand code	Grade levels
Addition	AD	0.10–8.90
Applications	AP	2.00–8.80
Decimals	DC	3.00–8.90
Division	DV	3.50–8.90
Equations	EQ	2.00–8.95
Fractions	FR	1.10–8.90
Geometry	GE	0.03–8.90
Measurement	ME	0.10–8.70
Multiplication	MU	2.50–8.90
Number concepts	NC	0.01–8.90
Probability and statistics	PR	7.00–8.90
Problem-solving strategies	PS	2.10–6.80
Science applications	SA	3.30–7.40
Speed games	SG	2.00–8.90
Subtraction	SU	0.60–8.90
Word problems	WP	0.50–8.90

Table 2. *Curriculum distribution by half grade of strands in MCS.*

Grade levels	AD	AP	DC	DV	EQ	FR	GE	ME	MU	NC	PR	PS	SA	SG	SU	WP
0.0	12	0	0	0	0	0	45	11	0	32	0	0	0	0	0	0
0.5	9	0	0	0	0	0	30	17	0	34	0	0	0	0	3	7
1.0	14	0	0	0	0	6	18	18	0	29	0	0	0	0	7	8
1.5	17	0	0	0	0	6	18	19	0	22	0	0	0	0	11	7
2.0	10	4	0	0	4	4	20	15	0	9	0	7	0	4	12	8
2.5	14	3	0	0	3	3	10	15	2	14	0	7	0	3	14	11
3.0	13	4	6	0	6	6	6	10	6	13	0	5	1	6	14	4
3.5	11	2	6	6	6	6	8	9	11	11	0	2	2	6	12	2
4.0	11	3	6	6	6	6	6	7	11	11	0	5	2	6	11	3
4.5	6	2	13	6	6	6	6	8	12	12	0	4	4	6	6	3
5.0	4	7	13	4	7	7	10	7	9	8	0	4	3	7	6	4
5.5	4	6	14	3	7	14	10	8	7	7	0	3	4	7	3	3
6.0	6	4	14	4	7	14	7	7	4	7	0	4	4	7	7	4
6.5	8	3	15	3	5	15	8	8	3	8	0	3	3	7	8	3
7.0	7	4	7	4	3	14	7	7	3	7	7	0	7	7	7	4
7.5	8	3	8	4	5	16	8	8	3	10	8	0	0	8	8	3
8.0	6	4	7	7	13	7	7	7	6	7	9	0	0	7	6	7
8.5	4	4	7	7	14	7	7	5	7	7	8	1	1	7	7	7

grade levels shown in Table 1 are for Grades 1–8 in standard American schools. The nature of all strands should be clear from their names, with the possible exception of speed games. This strand gives practice in improving the latency of response in basic facts of arithmetic, for example, the multiplication table for single-digit numbers. Continued improvements in speed of response long after no errors are made is familiar in many areas of performance. A simple mathematical model of these phenomena is proposed in Suppes, Groen, and Schlag-Rey (1966), but given the importance of such skills, more detailed and elaborate models are called for. An extensive account of performance latencies in arithmetic in terms of structural variables characterizing each exercise is given in Suppes and Morningstar (1972, ch. 5), but further detailed study of latency learning, as opposed to approximately asymptotic performance, is certainly still needed.

In Table 2 the curriculum frequency distribution is shown in half-grade intervals, taken from the 1993 *Teacher's Handbook for Math Concepts and Skills* (p. 19). Over the years we found that a fixed distribution for each grade was at too coarse a level, especially in the early grades. It is natural to ask what is the history of these detailed curriculum distributions, which are much more quantitative in nature than the usual curriculum guidelines used in schools. The analysis originally started with such guidelines. The first step in refining the guidelines was to follow in the footsteps of Edward Thorndike, probably the greatest educational psychologist in the first half of this century. Wanting to go beyond the bland generalities about what should be or what is the content of elementary mathematics textbooks, Thorndike made the original move of actually counting and classifying all the exercises in given texts. We did the same thing, but took several further steps as well. First, we converted the numbers to relative frequency or probability distributions, to be used in selecting exercises according to this distribution in our computer-based course.

The most important missing element is the structural analysis of prerequisites. For example, in teaching the algorithm of multiplication a number of prerequisite skills in addition are needed. Just a fragment of the prerequisite structure for the mathematics concepts and skills course is shown in Figure 1, which shows a conceptual dependency graph. The prerequisites start at the top and go down. This is for the very beginning of the course starting with number concepts at the beginning of the first grade. At the bottom of the dependency graph we have reached grade level 1.10 in addition, for example, 1.12 in geometry, 1.12 in measurement, and 1.12 in number concepts. What is important about this dependency graph is the way it can be used in the course. When a student is having difficulty on a particular concept or skill in a particular strand, it has turned out to be valuable to construct the minimum dependency graph for that skill. The most important aspect of review has been to review the

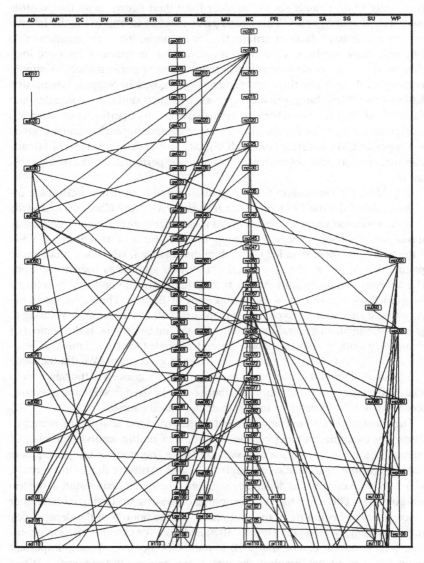

Figure 1. Partial dependency graph of prerequisites in MCS for first grade starting at the top with the number concept strand at grade placement 0.05.

154

Table 3. *Effect of prerequisite intervention on probability of an error.*

Skill	N	Before		After	
		1st Trial	2nd Trial	1st Trial	2nd Trial
AD 1.80	64	.76	.93	.48	.45
AD 2.15	208	.79	.95	.67	.56
AD 2.95	185	.68	.97	.44	.28
AD 3.80	211	.68	.91	.34	.33
DV 4.40	93	.83	.95	.46	.34
DV 5.40	129	.84	.96	.37	.36

prerequisites of a given skill in order to improve performance on it. Rather than simply having review on the skill itself, it turns out, when students are doing badly, it is more efficient to provide review of the prerequisites. In Table 3 we show some data on the reduction in probability of error for students who are having difficulty, after being given intervention by further work on pre-requisites of the given skill. The data shown on the table cover four different levels of addition and two different levels of division, mainly focused on the long-division algorithm. In looking at these data it is important to note that for students who were having real difficulties, the improvements did not reduce their errors to zero but in every case were substantial. Note that what is shown is the probability of an error on two trials before intervention and two trials after intervention on the given skill. In our own judgment the automatic provision for work on prerequisites is one of the most powerful and sophisticated features that can be used in a computer-based course. It is important to emphasize that a prerequisite structure of the kind shown can be important, even when there is not utter clarity and agreement on exactly what the prerequisite structure is, because the psychological analysis, as opposed to the purely mathemati-cal analysis, is less developed and less agreed upon, but has a robustness of its own.

Student distribution. However thorough the curriculum analysis that lies back of the curriculum distribution, individual student differences will necessarily lead to uneven progress for students across the range of concepts and skills in a given curriculum. One student will be much better at executing the standard algorithms of arithmetic than in solving word problems, and another student will be the reverse. So to keep the position of the student at approximately the same level of achievement in all the skills of a basic course in mathematics, a second, individual student distribution is introduced, for purposes of smoothing the actual distribution of grade level achievement across strands – the concepts

155

and skills are organized into homogeneous strands, one for fractions, one for word problems, and so forth.

We outline the basic setup, which depends on selecting two parameters. One is the threshold parameter θ for how far behind the average grade placement of student s in strand i must be to receive greater emphasis. The second is the parameter α for weighting the curriculum distribution $c(i)$, $\sum c(i) = 1$, and assigning weight $1 - \alpha$ to the distribution $k(i, s)$ of student s, defined next. Thus, at a given time t the actual distribution $d(i, s)$ used in selecting exercises from strand i for student s is defined as:

$$(1) \qquad d(i, s) = \alpha c(i) + (1 - \alpha)k(i, s),$$

with $0 \leqslant \alpha \leqslant 1$.

To define $k(i, s)$, let

$$(2) \qquad h(i, s) = \begin{cases} \bar{a}(s) - g(i, s) & \text{if } \bar{a}(s) \geqslant g(i, s) + \theta, \\ 0 & \text{otherwise,} \end{cases}$$

where $g(i, s)$ is, at the time t, the grade level achievement in strand i of student s, and $\bar{a}(s)$ is the weighted average grade level achievement of s at time t with the averaging being across strands weighted by the curriculum distribution $c(i)$. Finally,

$$(3) \qquad k(i, s) = \begin{cases} c(i) & \text{if } \sum_s h(i, s) = 0, \\ \dfrac{h(i, s)}{\sum_s h(i, s)} & \text{otherwise.} \end{cases}$$

The qualitative analogue of equation (1) is used by any observant intelligent human tutor. In this instance it is easy to implement something that probably does a better job in most cases.

Initial placement motion (IPM). The purpose of IPM is to move a student rapidly up or down in grade placement on the basis of performance in an initial sequence of sessions. The objective is to find the appropriate grade placement for the student to begin the course.

We begin with some preliminaries. Let $G_n, n \geqslant 0$, denote the student's average grade level at the end of session n, achieved under the standard curriculum motion, G_0 being the initial (entering) grade level. Then $\Delta_n = G_n - G_{n-1}$, $n \geqslant 1$, is the grade level increment (positive or zero) achieved during session n under the standard curriculum motion.

Let $Y_n, n \geqslant 1$, be the grade level increment (positive or negative) achieved at the end of session n under the IPM motion, which is defined explicitly later.

If $\mathbf{Z}_n$, $n \geqslant 1$, denotes the grade level at the end of IPM session n, we can write:

$$\mathbf{Z}_n = \mathbf{G}_0 + \sum_{i=1}^{n} \Delta_i + \sum_{i=1}^{n} \mathbf{Y}_i.$$

This means that the stochastic process $\{\mathbf{Z}_n; n \geqslant 1\}$, the grade level at the end of IPM session n, is the sum of the following processes:

(i) $\{\Delta_n; n \geqslant 1\}$, the curriculum process which is defined by the parameters of the standard curriculum motion
(ii) $\{\mathbf{Y}_n; n \geqslant 1\}$, the IPM process, defined as follows:

$$\mathbf{Y}_i = \begin{cases} \delta & \text{if } \dfrac{(TCOR)_i + \alpha_1}{(TATT)_i + \alpha_1 + \beta_1} \geqslant \gamma_1, \\[2ex] -\beta & \text{if } \dfrac{(TCOR)_i + \alpha_2}{(TATT)_i + \alpha_2 + \beta_2} \leqslant \gamma_2, \\[2ex] 0 & \text{otherwise,} \end{cases}$$

with α_1, α_2, β_1, β_2 positive real numbers, $\mathbf{Y}_1, \ldots, \mathbf{Y}_n$ independent random variables, $(TCOR)_i$ the total number of correct exercise in session i, and $(TATT)_i$ the total number of attempted exercises in session i. In order to complete the definition of the IPM process, the probability distribution of the random variables $\mathbf{Y}_1, \ldots, \mathbf{Y}_n$ must be given. These mathematical developments are not considered here.

Learning models of mastery. Our problem is to decide when performance on a given class of essentially equivalent exercises satisfies some criterion. In many situations it is assumed that the underlying process that is being sampled is stationary – at least in the mean. The first simple model we shall consider is of this type. A more realistic assumption in dealing with student behavior is that learning is occurring – both individually and in the mean – so that the process is not stationary. The second model is of this type. The third model is explained later.

Components (5) and (6) on forgetting models and tutorial intervention are discussed later.

3. LEARNING MODELS: THEORY

Although in the intended applications the number of possible student responses is usually large – and therefore the probability of guessing a correct answer is close to zero – we shall consider here only correct and incorrect responses. With this restriction:

$A_{0,n} = $ event of incorrect response on trial n,
$A_{1,n} = $ event of correct response on trial n,

157

x_n = possible sequence of correct and incorrect responses from trial 1 to n inclusive,

$q_n = P(A_0, n)$ = mean probability of an error on trial n,

$q_{x,n} = P(A_0, n \mid x_{n-1})$,

$q = q_1 = P(A_{0,1})$.

Also, $A_{0,n}$ and $A_{1,n}$ are the corresponding random variables. We shall also use $X_n = A_{0,n}$ as our most important random variable.

For simplicity we shall assume a fixed initial probability $q = P(A_{0,1})$ rather than a prior distribution on the unit interval for this probability. Given the extensive knowledge of this distribution, assuming that all the weight is on q is not unrealistic.

In Model I the assumptions are:

(i) $P(A_{0,n+1} \mid A_{0,n} x_{n-1}) = (1 - \omega) P(A_{0,n} \mid x_{n-1}) + \omega$,

(ii) $P(A_{0,n+1} \mid A_{1,n} x_{n-1}) = (1 - \omega) P(A_{0,n} \mid x_{n-1})$,

or equivalently, we have the single random-variable equation

(iii) $E(X_{n+1} \mid X_n, \ldots, X_1) = (1 - \omega) E(X_n \mid X_{n-1}, \ldots, X_1) + \omega X_n$.

We can easily prove the significant fact of stationarity of the mean probability $P(A_{0,n})$.

THEOREM 1 *In Model I, for all n, $P(A_{0,n}) = q_1$.*

Proof.

$$
\begin{aligned}
P(A_{0,n+1}) &= \sum_x [P(A_{0,n+1} \mid A_{0,n} x_{n-1}) P(A_{0,n} \mid x_{n-1}) P(x_{n-1}) \\
&\quad + P(A_{0,n+1} \mid A_{1,n} x_{n-1}) P(A_{1,n} \mid x_{n-1}) P(x_{n-1})] \\
&= \sum_x [(1 - \omega) P^2(A_{0,n} \mid x_{n-1}) + \omega P(A_{0,n} \mid x_{n-1}) \\
&\quad + (1 - \omega) P(A_{0,n} \mid x_{n-1})(1 - P(A_{0,n} \mid x_{n-1}))] P(x_{n-1}) \\
&= \sum_x P(A_{0,n} \mid x_{n-1}) P(x_{n-1}) \\
&= P(A_{0,n}).
\end{aligned}
$$

We have at once then

Corollary for Model I

(4) $$E(X_n) = q$$

(5) $$\mathrm{Var}(X_n) = q(1 - q).$$

158

In Model II we generalize Model I to unequal ωs on the assumption that learning is occurring during the trials, so we assume $\omega_1 < \omega_2$, and we replace (i) and (ii) by (i') and (ii'):

(i') $P(A_{0,n+1} \mid A_{0,n}x_{n-1}) = (1 - \omega_1)P(A_{0,n} \mid x_{n-1}) + \omega_1$
(ii') $P(A_{0,n+1} \mid A_{1,n}x_{n-1}) = (1 - \omega_2)P(A_{0,n} \mid x_{n-1}).$

To express results compactly, we define moments:

(6) $$V_{i,n} = \sum_x P^i(A_{0,n} \mid x_{n-1})P(x_{n-1}).$$

THEOREM 2 *In Model II,*

(7) $$V_{1,n+1} = (1 - (\omega_2 - \omega_1))V_{1,n} + (\omega_2 - \omega_1)V_{2,n}.$$

Proof. By the same methods used for Theorem 1,

$$\begin{aligned}
P(A_{0,n+1}) &= \sum_x [((1 - \omega_1)P(A_{0,n} \mid x_{n-1}) + \omega_1)P(A_{0,n} \mid x_{n-1}) \\
&\quad + (1 - \omega_2)P(A_{0,n} \mid x_{n-1})(1 - P(A_{0,n} \mid x_{n-1}))]P(x_{n-1}) \\
&= (1 - \omega_1)V_{2,n} + \omega_1 V_{1,n} + (1 - \omega_2)V_{1,n} - (1 - \omega_2)V_{2,n} \\
&= (1 - (\omega_2 - \omega_1))V_{1,n} + (\omega_2 - \omega_1)V_{2,n}.
\end{aligned}$$

In the case of both Models I and II the asymptotic behavior is well known (Karlin, 1953). All sample paths converge to 0 or 1, with the exact distribution depending on the initial distribution, and in the case of Model II, the relative values of ω_1 and ω_2. Of course, in the case of Model II, detailed computations are difficult. With the asymptotic dependence on initial conditions, neither process is ergodic.

Here is how either model would work computationally in practice. A student is exited upward from a class when $q_{x,n} < q^*$, where q^* is the normative threshold, for example, we might set $q^* = .15$, corresponding to a probability correct of .85. Notice that both models are noncommutative, and thus give greater weight to later responses. These models are derived from learning models that have been extensively studied. For pedagogical purposes in a computer environment they are computationally simple. The history of a sequence – 001100111, for example – is absorbed in the current $q_{x,n}$, and no other data need be kept, except possibly a count of the number of exercises.

Still more suitable is a third model that has both a parameter for individual paths, such as in Model I, or possibly two such parameters as in Model II, together with a uniform learning parameter α that acts constantly on each trial, since the student is always told the correct answer. For simplicity we shall

159

consider the two-parameter model using α and ω, which are assumed to lie in the open interval $(0, 1)$.

In Model III the assumptions are:

(iv) $P(A_{0,n+1} \mid A_{0,n}x_{n-1}) = (1 - \omega)\alpha P(A_{0,n} \mid x_{n-1}) + \alpha\omega$,

(v) $P(A_{0,n+1} \mid A_{1,n}x_{n-1}) = (1 - \omega)\alpha P(A_{0,n} \mid x_{n-1})$.

It is then easy to prove by the methods already used, or equivalently in terms of the random variable $\mathbf{X}_n$, the following:

(8) $\qquad E(\mathbf{X}_{n+1} \mid \mathbf{X}_n, \ldots, \mathbf{X}_1) = kE(\mathbf{X}_n \mid \mathbf{X}_{n-1}, \ldots, \mathbf{X}_1) + \alpha\omega\mathbf{X}_n$,

with $k = \alpha(1 - \omega)$.

PROPOSITION *Let* $\{\mathbf{X}_n, n \geqslant 1\}$ *be a learning process of type III. Then for each* $n \geqslant 1$ *we have:*

(9) $\qquad E(\mathbf{X}_{n+1} \mid \mathbf{X}_n, \ldots, \mathbf{X}_1) = k^n E(\mathbf{X}_1) + \alpha\omega \sum_{i=1}^{n} \mathbf{X}_i k^{n-i}$,

where $E(\mathbf{X}_1) = P(\mathbf{X}_1 = 1) = q$ *is the initial condition of the process.*

Proof. It follows from (8) by induction. From (8) with $n = 1$ we have:

(10) $\qquad E(\mathbf{X}_2 \mid \mathbf{X}_1) = P(\mathbf{X}_2 = 1 \mid \mathbf{X}_1) = kE(\mathbf{X}_1) + \alpha\omega\mathbf{X}_1$.

Thus, trial $n + 1$ of the process with initial condition $E(\mathbf{X}_1) = q$ can be viewed as the second trial of the same process with the observed value of $E(\mathbf{X}_n \mid \mathbf{X}_{n-1}, \ldots, \mathbf{X}_1)$ as initial condition. It is also clear from (9) that for each $n \geqslant 1$, the conditional expectation of $\mathbf{X}_{n+1}$ given $\mathbf{X}_i, i = 1, \ldots, n$, is a random linear function of $\mathbf{X}_i$ with coefficients $\alpha\omega k^{n-i}$.

We also have the recursion:

(11) $\qquad\qquad\qquad E(\mathbf{X}_{n+1}) = \alpha E(\mathbf{X}_n)$.

Proof.

$$E(\mathbf{X}_{n+1} \mid \mathbf{X}_n, \ldots, \mathbf{X}_1) = \alpha(1 - \omega)E(\mathbf{X}_n \mid \mathbf{X}_{n-1}, \ldots, \mathbf{X}_1) + \alpha\omega\mathbf{X}_n.$$

Then

$$\begin{aligned} E(\mathbf{X}_{n+1}) &= E(E(\mathbf{X}_{n+1} \mid \mathbf{X}_n, \ldots, \mathbf{X}_1)) \\ &= \alpha(1 - \omega)E(\mathbf{X}_n) + \alpha\omega E(\mathbf{X}_n) \\ &= \alpha E(\mathbf{X}_n). \end{aligned}$$

It follows at once from (11) that the mean learning curve is given by:

(12) $\qquad\qquad\qquad E(\mathbf{X}_{n+1}) = \alpha^n E(\mathbf{X}_1)$,

Similarly,

(13)
$$\text{VAR}(\mathbf{X}_n) = E(\mathbf{X}_n)(1 - E(\mathbf{X}_n)).$$

And we can then prove from (12) and (13):

(14)
$$\text{VAR}(\mathbf{X}_{n+1}) = \alpha^n E(\mathbf{X}_1)(1 - \alpha^n E(\mathbf{X}_1)).$$

We next consider the covariance recursion.

(15)
$$E(\mathbf{X}_{i+n+1}\mathbf{X}_i) = \alpha E(\mathbf{X}_{i+n}\mathbf{X}_i).$$

Proof.

$$
\begin{aligned}
E(\mathbf{X}_{i+n+1}\mathbf{X}_i \mid \mathbf{X}_{i+n}, \ldots, \mathbf{X}_1) &= \mathbf{X}_i E(\mathbf{X}_{i+n+1}\mathbf{X}_i \mid \mathbf{X}_{i+n}, \ldots, \mathbf{X}_1) \\
&= \mathbf{X}_i (k E(\mathbf{X}_{i+n} \mid \mathbf{X}_{i+n-1}, \ldots, \mathbf{X}_1) + \alpha\omega\mathbf{X}_{i+n}) \\
&= k E(\mathbf{X}_{i+n}\mathbf{X}_i \mid \mathbf{X}_{i+n-1}, \ldots, \mathbf{X}_1) + \alpha\omega\mathbf{X}_{i+n}\mathbf{X}_i.
\end{aligned}
$$

Then, taking expectation we have

$$
\begin{aligned}
E(\mathbf{X}_{i+n+1}\mathbf{X}_i) &= E(E(\mathbf{X}_{i+n+1}\mathbf{X}_i \mid \mathbf{X}_{i+n}, \ldots, \mathbf{X}_1)) \\
&= k E(\mathbf{X}_{i+n}\mathbf{X}_i) + \alpha\omega E(\mathbf{X}_{i+n}\mathbf{X}_i) \\
&= \alpha E(\mathbf{X}_{i+n}\mathbf{X}_i).
\end{aligned}
$$

By similar methods, we can easily show:

(16)
$$E(\mathbf{X}_{i+n+1}\mathbf{X}_i) = \alpha^n E(\mathbf{X}_{i+1}\mathbf{X}_i),$$

(17)
$$\text{COV}(\mathbf{X}_{i+n+1}\mathbf{X}_i) = \alpha^n \text{COV}(\mathbf{X}_{i+1}\mathbf{X}_i),$$

The expected number of errors in n trials is:

(18)
$$E\left(\sum_{i=1}^{n}\mathbf{X}_i\right) = E(\mathbf{X}_1)\frac{1 - \alpha^n}{1 - \alpha}.$$

In fact:

$$E\left(\sum_{i=1}^{n}\mathbf{X}_i\right) = \sum_{i=1}^{n} E(\mathbf{X}_i) = E(\mathbf{X}_1)\sum_{i=0}^{n-1}\alpha^i = E(\mathbf{X}_1)\frac{1 - \alpha^n}{1 - \alpha}$$

with

(19)
$$\lim_{\alpha \to 1} E\left(\sum_{i=1}^{n}\mathbf{X}_i\right) = nE(\mathbf{X}_i).$$

161

Furthermore,

$$\text{VAR}\left(\sum_{i=1}^{n} \mathbf{X}_i\right) = E(\mathbf{X}_1)\frac{1-\alpha^n}{1-\alpha}\left(1 - E(\mathbf{X}_1)\frac{1-\alpha^n}{1-\alpha}\right)$$

(20)
$$+ 2\sum_{i=1}^{n-1} \frac{1-\alpha^{n-i}}{1-\alpha} E(\mathbf{X}_{i+1}\mathbf{X}_i).$$

Learning following errors. We now prove some results about learning after errors are made. That learning is different after incorrect responses in comparison with correct responses is one of the most significant features of Model III.

PROPOSITION *For each* $n \geqslant 1$, $E(\mathbf{X}_{n+1} \mid \mathbf{X}_n, \ldots, \mathbf{X}_1) \leqslant \text{Max}\{q, \frac{\alpha\omega}{(1-k)}\}$.

Proof.

$$\text{Max}\, E(\mathbf{X}_{n+1} \mid \mathbf{X}_n, \ldots, \mathbf{X}_1) = E(\mathbf{X}_{n+1} \mid \mathbf{X}_i = 1, i = 1, \ldots, n)$$
$$= k^n\left(q - \frac{\alpha\omega}{1-k}\right) + \frac{\alpha\omega}{1-k}.$$

PROPOSITION *Let* $q \leqslant \alpha\omega/(1-k)$. *Then for each* $n \geqslant 1$ *we have:*

$$E(\mathbf{X}_{n+1} \mid \mathbf{X}_n, \ldots, \mathbf{X}_1) < E(\mathbf{X}_n \mid \mathbf{X}_{n-1}, \ldots, \mathbf{X}_1) \qquad \text{iff} \quad \mathbf{X}_n = 0.$$

Proof. The inequality can be written in the equivalent form:

$$E(\mathbf{X}_n \mid \mathbf{X}_{n-1}, \ldots, \mathbf{X}_1) > \frac{\alpha\omega}{1-k}\mathbf{X}_n.$$

The conclusion follows from the preceding proposition. These two propositions show that with $q \leqslant \alpha\omega/(1-k)$ and for each $n \geqslant 1$, $E(\mathbf{X}_{n+1} \mid \mathbf{X}_n, \ldots, \mathbf{X}_1)$ decreases if $\mathbf{X}_n = 0$ and increases if $\mathbf{X}_n = 1$. Thus, learning following errors will not occur. With $q > \alpha\omega/(1-k)$, let $q^* = q - [\alpha\omega/(1-k)]$. Then we can write:

$$E(\mathbf{X}_{n+1} \mid \mathbf{X}_n, \ldots, \mathbf{X}_1) = k^n q^* + \left(k^n\frac{\alpha\omega}{1-k} + \alpha\omega\sum_{i=1}^{n} k^{n-i}\mathbf{X}_i\right).$$

Following errors only we have:

$$E(\mathbf{X}_{n+1} \mid \mathbf{X}_i = 1, i = 1, \ldots, n) = k^n q^* + \frac{\alpha\omega}{1-k}.$$

Thus, learning following errors only will occur if and only if $q > \alpha\omega/(1-k)$ and its magnitude will be at most q^*. In general, for each $n \geqslant 1$ the realizations

162

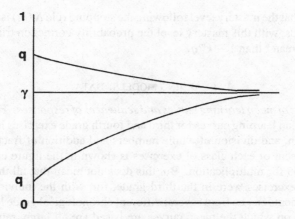

Figure 2. The graph shows the Model III learning curves for a sequence of errors only; the two cases depend on whether the initial q is less than or greater than γ.

of $E(\mathbf{X}_{n+1} \mid \mathbf{X}_n, \ldots, \mathbf{X}_1)$ satisfy the inequality

$$k^n q \leqslant E(\mathbf{X}_{n+1} \mid \mathbf{X}_n, \ldots, \mathbf{X}_1) \leqslant k^n q + \alpha\omega\frac{1 - k^n}{1 - k}.$$

Figure 2 shows the importance of the bound $\gamma = \alpha\omega/(1 - k)$.

Sequential analysis. We now turn to joint probabilities such as $P(\mathbf{X}_1 = x_1,$ $\mathbf{X}_2 = x_2, \mathbf{X}_3 = x_3, \mathbf{X}_4 = x_4)$, where $x_i = 0$ or 1. First,

$$\begin{aligned} P(\mathbf{X}_1, &\ldots, \mathbf{X}_{n+1}) \\ &= P(\mathbf{X}_1)P(\mathbf{X}_2 \mid \mathbf{X}_1)P(\mathbf{X}_3 \mid \mathbf{X}_1, \mathbf{X}_2) \cdots P(\mathbf{X}_{n+1} \mid \mathbf{X}_1, \ldots, \mathbf{X}_n). \end{aligned}$$

We have immediately $P(\mathbf{X}_1 = 1) = q$, $P(\mathbf{X}_1 = 0) = 1 - q$ and with $i \geqslant 1$:

$$P(\mathbf{X}_{i+1} = 1 \mid \mathbf{X}_1, \ldots, \mathbf{X}_i) = kP(\mathbf{X}_i = 1 \mid \mathbf{X}_1, \ldots, \mathbf{X}_{i-1}) + \alpha\omega\mathbf{X}_i.$$

In MCS the following mastery-stopping rule was used for moving to the next concept in a strand, with different parameter values set for different concepts. In fact, the concepts were, for this purpose, put in one of six classes. We shall not go into the details of parameter selection for these six classes.

DEFINITION *A mastery-stopping rule adapted to the process $\{\mathbf{X}_n, n \geqslant 1\}$ with threshold $t = k^m q, m \geqslant 1$, is a random index of the form:*

$$(21) \quad N(m) = \begin{cases} \inf\{n : E(\mathbf{X}_{n+1} \mid \mathbf{X}_1, \ldots, \mathbf{X}_n) \leqslant k^m q\} & \textit{if this set is} \\ & \quad \textit{nonempty,} \\ +\infty & \textit{otherwise.} \end{cases}$$

163

We will say that the mastery level following the stopping rule $N(m)$ is $(1-k^m q)$. In other words, with this mastery level the probability correct on trial $N(m)$ is equal to or greater than $1 - k^m q$.

4. LEARNING MODELS: DATA

Data analysis of mean learning curves and sequences of responses. Figures 3–6 show four mean learning curves for third and fourth grade exercises in addition, multiplication, and division of whole numbers and addition of fractions. The grade placement of each class of exercises is shown in the figure caption, for example, 3.55 for multiplication. But this does not mean that all the students doing these exercises were in the third grade, for, with the individualization possible, students can be from several different chronological grade levels. The sample sizes on which the mean curves are based are all large, ranging from 611 to 1,283 students. The students do not come from one school and certainly are not in any well-defined experimental condition. On the other hand, all of the students were working in a computer laboratory in an elementary school run by a proctor, so there was supervision of a general sort of the work by the students, especially in terms of schedule and general attention to task. For each

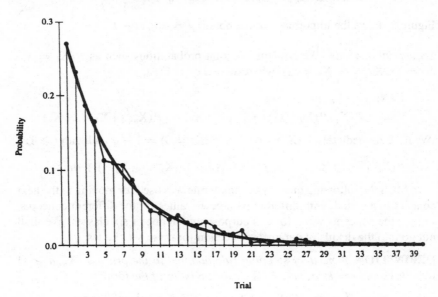

Figure 3. Mean learning curve for addition strand of MCS at grade level 3.10; the sample size is 611 students, $q = 0.269$, $\alpha = 0.840$, standard error of estimate (s.e.e.) = 0.0059, mean absolute deviation (m.a.d.) = 0.0041, max. dev. = 0.0209.

164

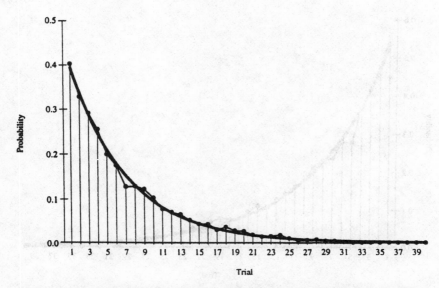

Figure 4. Mean learning curve for subtraction strand of MCS at grade level 4.10; the sample size is 719 students, $q = 0.394$, $\alpha = 0.854$, s.e.e. = 0.0062, m.a.d. = 0.0042, max. dev. = 0.0254.

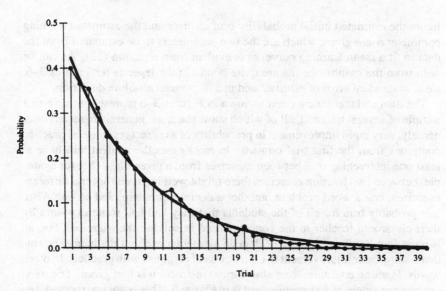

Figure 5. Mean learning curve for multiplication strand of MCS at grade level 3.55; the sample size is 861 students, $q = 0.423$, $\alpha = 0.884$, s.e.e = 0.0103, m.a.d. = 0.0087, max. dev. = 0.0237.

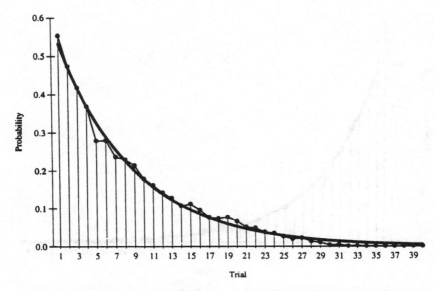

Figure 6. Mean learning curve for fraction strand of MCS at grade level 3.90; the sample size is 1,283 students, $q = 0.533$, $\alpha = 0.885$, s.e.e. $= 0.0155$, m.a.d. $= 0.0082$, max. dev. $= 0.0478$.

figure the estimated initial probability q of an error and the estimated learning parameter α are given, which are the two parameters to be estimated from the data to fit a mean learning curve, as is evident from equation (12). As can be seen from the graphs, the fits are quite good. In the legends for Figures 3–6, s.e.e. = standard error of estimate and m.a.d. = mean absolute deviation.

The data and theoretical curves shown in Figures 3–6 represent four from a sample of several hundred, all of which show the same general characteristics, namely, very rapid improvement in probability of a correct response as practice continues from the first trial onward. In most cases the student will have at least one intervening trial between exercises from a given class. So, for example, between two fraction exercises there might well intervene several different exercises, one a word problem, another a decimal problem, and so on. Also, it is probably true for all of the students that they had had some exposure by their classroom teacher to the concepts used in solving the exercises, but, as is quite familiar from decades of data on elementary school mathematics, students show clear improvement in correctness of response with practice. In other words, learning continues long after formal instruction is first given. The most dramatic example of an improvement is in Figure 6. This is not unexpected, because understanding and manipulation of fractions are among the most difficult concepts for elementary school students to master in the standard curriculum.

166

In Figures 7–10 data from the same four classes of exercise are analyzed in terms of the more demanding requirement on the learning model to fit the sequential data. For learning theorists the stringency of the test to fit such sequential data with only three parameters is well known. The data in each of the figures have twelve degrees of freedom because of the three parameters estimated from the data. The largest χ^2 is for the multiplication exercises, but even here the χ^2 is not significant at the 0.10 level, which is indicative of the good fits. Moreover, parameters of q and α were estimated only from the mean learning curve data. Only the parameter ω was directly estimated from the sequential data. So the test of fit is a stringent one.

Modification of Model III. In examining the fit of a number of mean learning curves, we found that by changing the time scale monotonically, the fit could sometimes be improved. In particular, we modified the mean learning curve given in (12) by replacing the exponent n of α, where, of course, n is the number of trials by n^β, so that we may write the equation for mean learning as:

$$(22) \qquad q_n = \alpha^{n^\beta} q.$$

In Figures 11 and 12, we show comparative results for one class of exercise in division at the third grade level. In Figure 11, the value of β is 1.0, and in Figure 12, β is 1.258. The improvement in fit from using the second β is

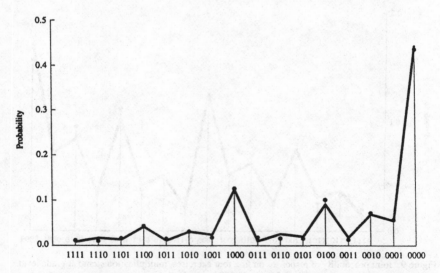

Figure 7. Joint probability of responses on first four exercises, addition strand at grade level 3.10; $\omega = 0.112$, $\chi^2 = 14.2$. Observed data shown by black dots.

167

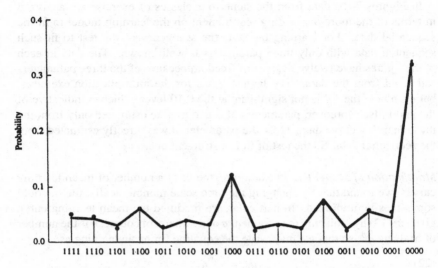

Figure 8. Joint probability of responses on first four exercises, subtraction strand at grade level 4.10; $\omega = 0.219$, $\chi^2 = 4.9$.

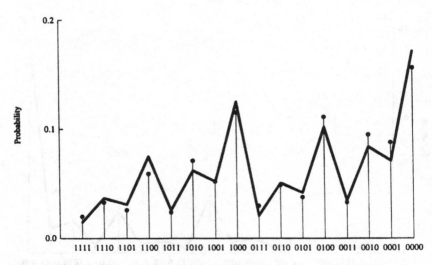

Figure 9. Joint probability of responses on first four exercises, multiplication strand at grade level 3.55; $\omega = 0.000$, $\chi^2 = 16.9$.

168

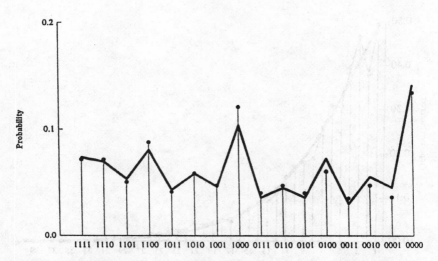

Figure 10. Joint probability of responses on first four exercises, fraction strand at grade level 3.90; $\omega = 0.126$, $\chi^2 = 14.7$.

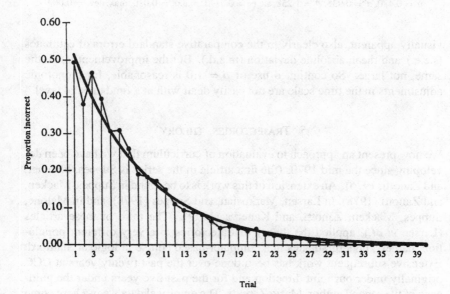

Figure 11. Mean learning curve for division strand of MCS at grade level 3.80; the sample size is 653 students, $q = 0.516$, $\alpha = 0.878$, s.e.e. $= 0.021$, m.a.d. $= 0.013$, max. dev. $= 0.074$.

169

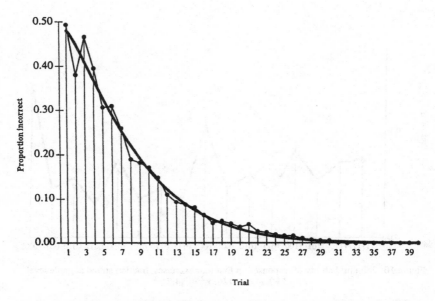

Figure 12. Same data for division as Figure 11, but additional parameter β of equation (22): $q = 0.480$, $\alpha = 0.935$, $\beta = 1.258$, s.e.e. $= 0.0172$, m.a.d. $= 0.010$, max. dev. $= 0.069$.

visually apparent, also clearly in the comparative standard errors of estimates (s.e.e.) and mean absolute deviation (m.a.d.). But the improvement is, all the same, not large. So continued use of $\beta = 1.0$ is reasonable, for monotonic adjustments in the time scale are not easily dealt with at a fundamental level.

5. TRAJECTORIES: THEORY

We now present an approach to evaluation of curriculum that we have been developing since the mid 1970s (the first article in the series is Suppes, Fletcher, and Zanotti, 1976). An extension of this work is to be found in Suppes, Macken, and Zanotti (1978), in Larsen, Markosian, and Suppes (1978), and in Malone, Suppes, Macken, Zanotti, and Kanerva (1979). The third of these articles (Larsen *et al.*) applied the theory of trajectories to a very different population, namely, undergraduates in a computer-based course in logic at Stanford. Extensive subsequent work has been done over the past twenty years at CCC originally under our joint direction and for the past five years under the guidance of the second author, Mario Zanotti. The empirical data we use here come from the extensive work at CCC. We turn now to a general introduction to this aspect of curriculum.

170

Many of us who have engaged in curriculum reform efforts have been dissatisfied with the wait-and-see approach required when classical evaluation of a new curriculum is used. We have in mind evaluation by comparing pretests and posttests, with an analysis of posttest grade placement distributions as a function of pretest distribution and exposure in some form to the new curriculum.

In line with approaches used in other parts of science, it is natural to ask if a more predictive-control approach could be used and made an integral part of the curriculum to ensure greater benefits, especially for the students not close to the mean performance. The approach discussed in this chapter is aimed precisely at this question. The strategy is to develop a theory of prediction for individual student progress through the curriculum, to use this predictive mechanism as a means of control by regulating the amount of time spent on the curriculum by a given student, and thereby to achieve set objectives for the grade-placement gains of the student. Such an approach also calls for individualization in the objectives of a course, for it is unrealistic to expect all students to make the same gains in the same amount of time, or to expect that the slowest students can cover as much material as the best students simply by spending additional time. Consequently, even with a differential approach to the amount of time each student may spend in the curriculum, it is still not reasonable to impose a uniform concept of grade placement gain on all students.

Another important feature of our approach to the prediction of student progress is to separate the global features of the curriculum from the global individual parameters characterized for the individual student by a simple differential equation. In many respects, the estimation of the global individual parameters corresponds to the fixing of boundary conditions in the solution of differential equations in physics. In our case, the boundary conditions correspond to the characteristics of the individual student and the differential equation itself to the structure of the curriculum.

As we have already emphasized, our analysis is aimed at the global performance of the student. The fact that we are considering only global progress, and not performance on individual exercises, makes it possible for us to state general axioms about information processing from which we may derive the basic stochastic differential equation that we believe is characteristic of many different curriculums, especially curriculums that are tightly articulated and organized in their development. Certainly this is a characteristic of the computer-assisted instruction in elementary mathematics considered here.

Let us assume, as already indicated, that the student is progressing individually through a course, and let $I(t)$ be the total information presented to the student up to time t. We say here *total information* but we could also give a formulation in terms of *concepts* or *skills* and think of the development procedurally rather than declaratively. Let $y(t)$ be the student's course position at time t. Note that for simplification of notation we have omitted a subscript for a particular student. It is understood that the notation used here applies only to an individual student not to averages of students. The stochastic averaging involved is averaging over the variety of skills or information presented to the student and refer to the student's mean position and mean information. We shall not make these stochastic assumptions explicit any further, because only the mean theory for the individual student will be developed here.

The first assumption is that the process is additive using for simplicity of concept a discrete-time variable n. We may write the additivity assumption as follows:

(23)   Additive: $I(n) - I(n-1) = \alpha$,

which we can then express in terms of a derivative for continuous time as:

$$(24) \qquad \frac{\dot{I}(t)}{I(t)} = \frac{1}{t}.$$

We then make the second strong assumption that the position in the course is proportional to the information introduced, that is,

$$(25) \qquad y(t) \approx I(t).$$

Combining (24) and (25) we have:

$$(26) \qquad \frac{\dot{y}(t)}{y(t)} \approx \frac{1}{t},$$

which we then integrate to obtain:

$$(27) \qquad \ln y(t) = k \ln t + \ln b,$$

which we may express so that $y(t)$ is a power function of t:

$$(28) \qquad y(t) = bt^k.$$

We use parameters b and k to fit the shape of a student trajectory. We add another parameter a to use in estimating the student's starting, grade placement position, so that our final equation is

$$(29) \qquad y(t) = bt^k + a.$$

6. TRAJECTORIES: DATA

We now turn to the analysis of a rather large number of student trajectories based on data collected in CCC's courses in elementary mathematics and reading. We examine data on 1,485 students collected during the 1992–1993 school year from students in various parts of the United States.

172

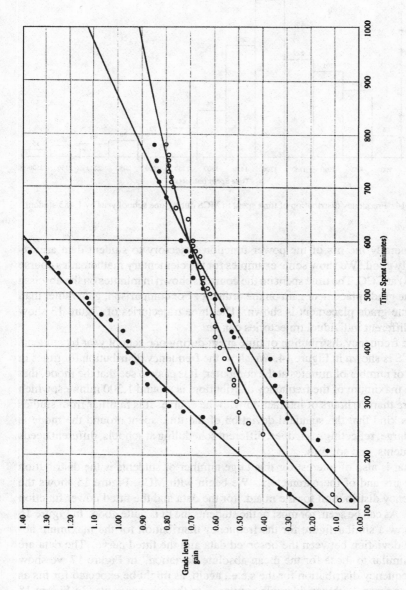

Figure 13. Example of three student trajectories.

173

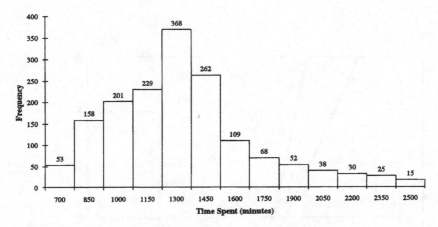

Figure 14. Frequency distribution of time spent in MCS during one school year by 1,485 students.

Generally the fits of the power function trajectory to student data are extremely good. We show some examples for the elementary mathematics course (MCS) at CCC. The time spent on the course is shown in minutes on the abscissa and the grade placement gain on the ordinate. For comparison, gain rather than absolute grade placement is shown. The three trajectories of Figure 13 show how different individual trajectories can be.

The frequency distribution of time spent during one school year by students in MCS is shown in Figure 14. We show the frequency distribution for times in terms of number of minutes at the computer. It is easy to see that the mode, that is, the maximum of the frequency distribution, is around 1,300 minutes, which is more than 20 hours of interaction with the course. It is familiar from studies of this kind that the standard deviation of the time spent around the mode is quite large, reflecting, as it does, different scheduling at schools, different needs of students, and so forth.

What is also of interest for this large number of students is the distribution of the fits and of the parameters. We begin with MCS. Figure 15 shows the frequency distribution for the m.a.d. for the data and the fitted power function curve. As can be seen, for most of the students the fit is quite good. In Figure 16 we show a similar figure for the frequency distribution for the maximum absolute deviation between the observed data and the fitted curve. The data are very similar to those for the mean absolute deviation. In Figure 17 we show the frequency distribution for the s.e.e., again, as might be expected for fits as good as these, with considerable similarity to the other two fits. In Figure 18 we show the frequency distribution for parameter a. This mainly just shows

174

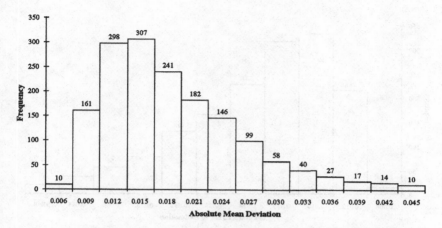

Figure 15. MCS frequency distribution of the m.a.d. for the fit of the 1,485 power function curves to the data.

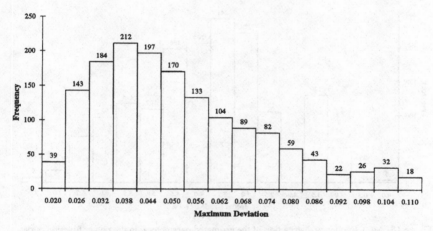

Figure 16. MCS frequency distribution of the m.a.d. for the fit of the 1,485 power function curves to the data.

at what grade level the students began. As can be seen, the students in the sample range from beginning first graders to seventh graders with the mode in the third grade. Figure 19 shows the frequency distribution for the parameter k, the exponent in the power function. For estimates based on a large number of data points and for any extended time, we would be surprised to have estimates for $k > 1$. As can be seen, there are more than three hundred in our sample,

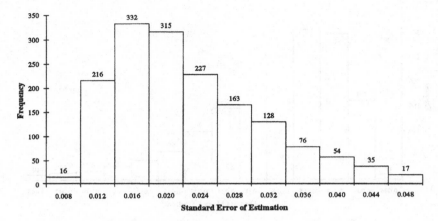

Figure 17. MCS frequency distribution of the s.e.e. for the fit of the 1,485 power function curves to the data.

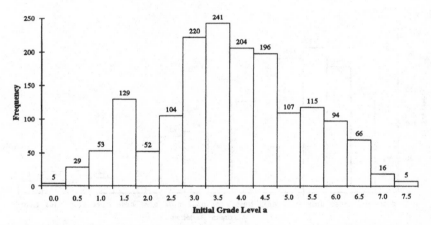

Figure 18. MCS frequency distribution of the estimated parameter a in the power function curve for 1,485 students.

and these estimates arise when we are working from an initial segment when students often show an accelerated rate of learning and therefore have an estimate of $k > 1$. We would not expect these large estimates of k to extend over a substantial part of the school year. What is important about the figure is to show how great the range of k is. If we exclude the two tails the range is from 0.4 to 1.4, a difference that leads to very considerable differences in rates of progress corresponding to rates of learning. We emphasize, on the other hand, that the

176

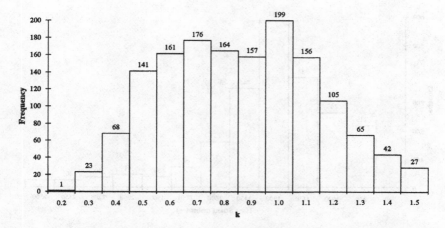

Figure 19. MCS frequency distribution of the estimated parameter k in the power function for 1,485 students.

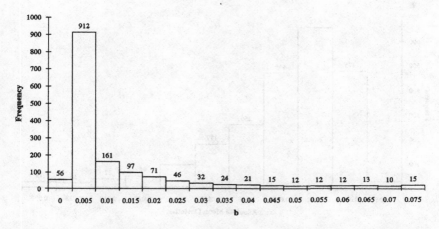

Figure 20. MCS frequency distribution of the parameter b in the power function for 1,485 students.

existence of such large individual differences in the student population is to be found in almost every other kind of estimation of achievement. But ordinarily these measures of achievement are for static cross-sectional tests, not for rates of progress during a period of at least several months.

In Figure 20 we show the distribution of the parameter b, which , unlike k, has a relatively small standard deviation around the mean value of approximately 0.005. Again, it is the exponent k that reflects strongly the individual

177

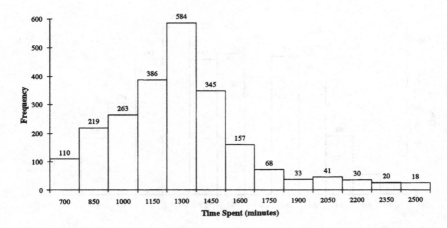

Figure 21. Frequency distribution of time spent in the reading course RW during one school year by 1,485 students.

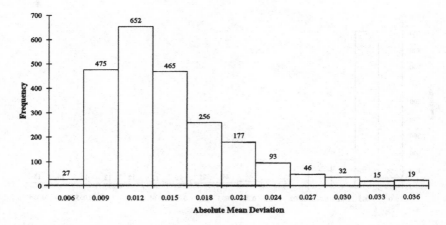

Figure 22. RW frequency of m.a.d. of the fit of the 1,485 power function curves to the data.

differences, not b, and not a, which corresponds just to where the student started and is highly correlated with age. The frequency distribution shown in Figure 21 for time spent in the reading course Reading Workshop (RW) is similar to the corresponding Figure 14 for MCS. And again the mode is about 1,300 minutes. This similarity in modes is not accidental. The students ordinarily spend about 10 minutes daily in each course. What the data show is that the modal student is spending about 130 days of the school year working on the two courses, usually in immediate daily sequence. Figure 22 shows the frequency

178

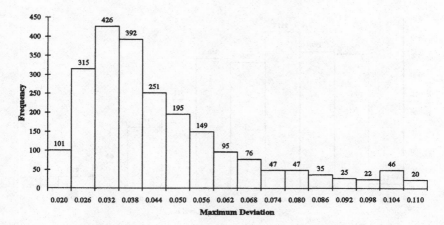

Figure 23. RW frequency distribution of the maximum absolute deviation for the fit of the 1,485 power function curves to the data.

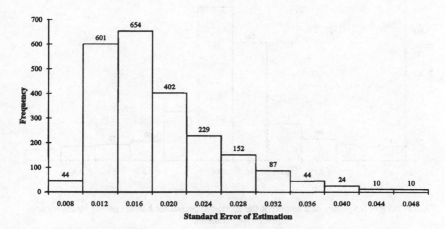

Figure 24. RW frequency distribution of the s.e.e. for the fit of the 1,485 power function curves to the data.

distribution for m.a.d. corresponding to Figure 15 for MCS, Figure 23 the frequency distribution for the maximum absolute deviation corresponding to Figure 16 for MCS, and Figure 24 the frequency distribution of the s.e.e. for RW corresponding to Figure 17 for MCS. Generally speaking, the data are quite similar to those for MCS.

179

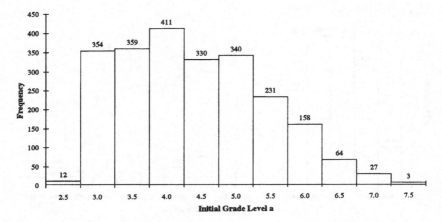

Figure 25. RW frequency distribution of the estimated parameter a in the power function curve for 1,485 students.

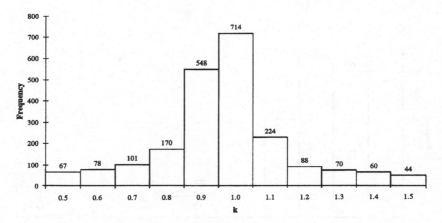

Figure 26. RW frequency distribution of the estimated parameter k in the power function curve for 1,485 students.

The similarity to MCS is also to be found in the distribution of the parameter a for RW, as shown in Figure 25. On the other hand, the distribution of the exponential parameter k is less spread out in the case of RW than in the case of MCS, as may be seen from Figure 26 in comparison with Figure 19. We have conjectures for this rather striking difference in the two courses, but we are not certain of their correctness. Finally, in Figure 27 we show the distribution for RW of the parameter b, which again is similar to that for MCS (Figure 20) but has a smaller value for the mode and a slightly larger standard deviation.

180

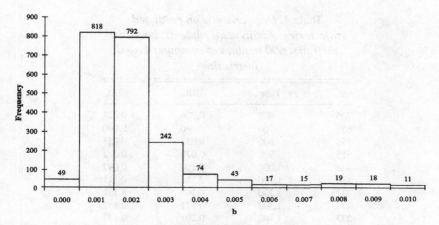

Figure 27. RW frequency distribution of the estimated parameter b in the power function curve for 1,485 students.

7. TRAJECTORIES: PREDICTION AND INTERVENTION

It is a familiar mathematical fact that a power function, like an exponential function, is quite unstable, and therefore simple predictions based on the power function are often not going to work very well, even though the systematic description of given data is very accurate. We have taken account of these fundamental phenomena in two ways. First, we emphasize that the task is not one of simply forecasting or predicting outcomes after many months but, rather, the combined task of predicting and intervening so we make relatively short-term forecasts and then make further corrections after a short period of time in order to achieve some longer-term objective. Second, we do not use prediction and intervention necessarily only for a student's individual trajectory. We have found that even in the case of predicting and intervening over relatively short periods, it is still better in most instances to use the average trajectory over a population of students as the "base" from which to predict. The details of all this will not be pursued here. Elaborate developments and the results of applications are to be found in a sequence of CCC Technical Notes and Memoranda circulated to the school systems and used by them for a number of years. We present only one set of data here to give a detailed sense of how the accuracy of predictions falls off the further out the prediction is. In the first column of Table 4 we show the number of students on which the calculation is based. We calculated trajectories in MCS according to the theory of Section 5, and we used the entire student trajectory as baseline data to compare with predictions. The predictions for all rows of the table were based only on the first

181

Table 4. *Data analysis on predicted trajectories of individual students, based on their first 600 minutes of computer-based instruction.*

N	Time	Diff.	S.D.
705	600	0.026	0.027
659	700	0.040	0.040
595	800	0.058	0.055
515	900	0.078	0.072
436	1,000	0.100	0.092
384	1,100	0.121	0.111
331	1,200	0.145	0.132
278	1,300	0.171	0.152
233	1,400	0.201	0.177
196	1,500	0.237	0.209
170	1,600	0.271	0.237
137	1,700	0.297	0.270
104	1,800	0.325	0.295
88	1,900	0.344	0.331
72	2,000	0.377	0.345

600 minutes. In each row we are comparing the predicted grade placement after x minutes for the students who had during the school year at least that many minutes of computer time, with the grade placement inferred from the trajectory fitted to all the data. We used the trajectory fitted to all the data to provide the data point to compare with prediction, because we did not have direct observations of grade placement every 100 minutes. As is evident, we show cumulative time in column 2 of Table 4, and the mean absolute difference between the predicted and data-fitted grade placements in column 3. In column 4 we show the corresponding standard deviation for the predicted and data-fitted grade placements. We emphasize again that the predictions for all rows were based on the data-fitted trajectory for the first 600 minutes, so both columns 2 and 3 show, as expected, monotonically increasing deviations between prediction and data fit as the point of prediction of grade placement is further away from the initial 600 minutes.

8. MOVEMENT AND MASTERY: NEW VERSION

In this section we turn to our conceptualization of student movement and mastery criteria for a new course at Stanford that includes not only review and practice, but also short "audiovisual lectures" on each major concept as it is introduced. In other words, we now turn to the problem of movement and mastery

182

in what is meant to be a self-contained course in elementary mathematics for grades K–8. In the context of the present analysis we shall not really discuss the content of the course, which is based on an extensive revision of the first author's elementary mathematics textbook series *Sets and Numbers* (1963). Here we only want to concentrate on drawing lessons from our experience with the mathematics concepts and skills course at CCC and related courses, as well as our work in other areas on learning, to design a new and, we believe, still more dynamic approach to problems of movement and mastery.

We begin with the classification of each trial, or, if you will, each exercise, into one of eight types. This partition of exercises is, of course, not the most refined one, but we believe it constitutes a sufficiently fine partition for purposes of making conceptual distinctions about learning and forgetting, and at the same time is not so fine but that we can seriously track individual students at each stage in terms of a parameterized version of these eight kinds of learning and forgetting. Notice that Learning Model III is incorporated quite directly into this classification. The model of forgetting goes back to earlier work (Suppes, 1964). As far as we know, the explicit introduction of models of forgetting in the analysis and management of computer-based curriculum has not previously taken place, even though in many kinds of instruction intuitive and implicit assumptions about forgetting are made, as for example in the various regimes of review and practice that are set up.

I. Movement and computation of q_n

Beginning with the first learning trial on a given concept C_i, any later trial on any task or concept has one of the following classifications with respect to concept C_i, as long as C_i is in the active set A (see III. 5 and V. 1). The model and parameters of change of q_n for each of the eight types are shown after the description. Remember, q_n is the probability of an error on concept C_i on trial n – we omit for simplicity the subscript i on q, which is understood.

1. A learning trial on C_i with a correct response
$$q_{n+1} = (1 - \omega)\alpha q_n.$$
2. A learning trial on C_i with a wrong response
$$q_{n+1} = (1 - \omega)\alpha q_n + \alpha\omega.$$
3. A review trial on C_i with a correct response
$$q_{n+1} = (1 - \omega)\beta q_n.$$
4. A review trial on C_i with a wrong response
$$q_{n+1} = (1 - \omega)\beta q_n + \beta\omega.$$

5. A "same strand" trial on a concept that has C_i as a prerequisite and a correct response

$$q_{n+1} = \alpha q_n.$$

6. A "same strand" trial on a concept that does not have C_i as a prerequisite, but a correct response.

$$q_{n+1} = \delta q_n.$$

7. For a "same strand" trial, if the response is incorrect,

$$q_{n+1} = q_n.$$

8. An "other" trial on a strand to which C_i does not belong

$$q_{n+1} = (1 - \epsilon)q_n + \epsilon.$$

II. Criterion of movement

1. Grade placement of concepts. Each concept C_i of each strand i has a grade placement $GP(C_i)$. Example, $GP(C_i) = 4.50$. The integer 4 is the grade and .50 is the position of this concept in the fourth grade.
2. At any time, a student has a current grade placement in each strand, which we express as $GP(i, s)$. This current $GP(i, s)$ shows what concept the student is working on in strand i.
3. Let

$$GP(i^*, s) = \max_i GP(i, s),$$

that is, i^* is the strand in which student s currently has the highest grade placement. There can be several strands at the same position, so the max does not have to be unique.
4. For each strand i, let $N(i, s)$ be the number of concepts between $GP(i, s)$ and $GP(i^*, s)$.
   (a) If $GP(i^*, s) - GP(i, s) > 0$, take the number to be at least 1.
   (b) Because the number of concepts varies in different strands, the count of number of concepts between the two grade placements $GP(i^*, s)$ and $GP(i, s)$ is a better measure of curriculum to be covered in strand i than is the numerical difference $GP(i^*, s) - GP(i, s)$.
5. We use $N(i, s)$ to compute a current probability distribution of curriculum emphasis for student s – as can be seen, this probability distribution varies from one student to another, and from one time to another for the same student.
   (a) $N(s) = \sum_i N(i, s).$

(b) $p(i, s) = N(i, s)/N(s)$ if $N(s) > 0$.

Use the rational fractions $p(i, s)$ to choose probabilistically the next strand i, sampling with replacement.

(c) If $N(s) = 0$, that is, all strands have the same $GP(i, s)$, use the uniform distribution, that is, weight all strands equally, in choosing the next strand.

(d) Use the uniform distribution initially when all strands have the same grade placement.

6. Choice of learning L or review R after choice of strand i.

(a) If no concepts of strand i are in active set A such that $q_{i,n} > q^*$, go to next new concept C_i.

(b) Let $j = |A_i|$ be cardinality of concepts in A from strand i such that

$$q_{i,n} > q^*,$$

where q^* is parameter for review. By assumption $j > 0$.

(c) Choose R with probability $j/(j + 1)$.

(d) Choose L then with probability $1 - [j/(j + 1)]$.

(e) Having chosen R, review concept C_i in A that has maximum $q_{i,n}$. If a tie, choose randomly among the maximum set.

(f) If L is chosen, then, as already stated, go to next concept C_i in strand i for student s.

III. Criterion of learning

1. On initial presentation of a concept, choose exercises by a slightly modified Fibonacci sequence, for example,

Differences: 1,2,3,5,8,13
Exercise no.: 1,3,6,11,19,32.

2. If mastery does not result from a Fibonacci sequence on first block, then run a Fibonacci sequence again on the remaining sequence of exercises not used the first time.

3. On any learning trial, if

$$q_n < m,$$

then the mastery criterion, as determined by the parameter m, is satisfied.

4. If the mastery criterion is not met, continue learning trials until no trials are left.

5. Whether either 3 or 4 obtains, after learning trials end, put C_i in active set A.

IV. Choice of review exercises

1. When a concept C_i is selected for review from active set A, randomly choose two exercises from the learning exercises following this concept.
2. If both exercises have wrong response, go to lecture on C_i and then randomly choose two other exercises from the learning exercises following this concept. But go as part of review of lecture L on C_i a maximum of λ times.
3. Otherwise go back to Section II and choose a strand i, then L or R as before.

V. Removal of concept from active list A

1. If C_i from strand i is in A, remove from A if and only if

$$GP(i, s) - GP(C_i) > \gamma,$$

where

$$GP(i, s) = \text{grade placement of student } s \text{ on strand } i$$
$$GP(C_i) = \text{grade placement of concept } C_i \text{ in the curriculum.}$$

We ordinarily take $\gamma = 0.5$, a half-year in grade placement.

VI. Updated $GP(i, s)$

Every time a new concept C_i is introduced to a student, $GP(i, s)$ is then updated to the position of C_i, that is,

$$GP(i, s) = GP(C_i).$$

(This remains true even when a concept C_i is failed in the present version.)

VII. Initial values of parameters for first experiment

$$q = .5, \alpha = .9, \omega = .1, m = .23, \beta = .8$$
$$\epsilon = .001, \delta_1 = \delta_2 = .99, q^* = .3, \gamma = .5, \lambda = 2.$$

9. CONCLUDING REMARKS

A quick inspection shows that this new version of mastery learning differs in significant ways from what is described in earlier sections and has been used extensively in CCC courses. Obviously some ingredients of the new system are based on more recent scientific results than others. Perhaps the most important feature of which this is true is the forgetting model embodied in the eighth type of movement under I at the beginning of Section 8, that is, the equation

$$q_{n+1} = (1 - \epsilon)q_n + \epsilon,$$

which applies to a trial presenting an exercise not belonging to the strand to which C_i belongs. This equation provides a very simple model of forgetting – the probability of error increases slightly when other exercises from another strand are given. The model is, of course, much too simple. We would naturally expect forgetting to vary significantly as a function of both student and concept. As new data accumulate, we hope to at least introduce a parameter for each student and each strand.

A critical dynamic feature of the new version is the active set A of concepts to be reviewed, especially the subset A^* of A whose error rates are above the criterion q^* according to the model computations used. Computing the expected size of A or A^* as a function of the model parameters is too complex to be feasible, but we have simulated the system with the parameters shown in VII.

We believe that the dynamic features we have built into the review process should accommodate a broad range of data and models concerned with forgetting. We also emphasize that it is most desirable that the models used go beyond the phenomenological data of forgetting to the causes of the phenomena, causes that have been much studied in the experimental literatures of psychology over many years, but hardly used at all in quantitative approaches to curriculum organization.

The third, and final, critical aspect of the new version we consider is the method of tutorial intervention. In the version being implemented and tested at Stanford in the spring of 1995, we have restricted tutorial intervention to that given in IV. 2, which is to return the student to the instructional lecture on a concept when difficulties arise in doing the relevant review exercises. In the future, we plan to add new, brief tutorial lectures on concepts that seem to require it, as judged by the performance of the students. Second, we plan to add brief tutorial interventions for particular exercises that students get wrong with high probability and for reasons we can diagnose.

These remarks touch on only a few of the many ideas we have for improvement. Students and their parents are already suggesting many other features they would find attractive. It is part of our Bayesian viewpoint outlined at the beginning of this article that there is no end to such improvements. It is particularly important for us to stress the importance of revisions based on systematic theory and data, even though we do not expect the results in any sense to be able fully to dictate the changes. A Bayesian place for intuitive judgment must remain, even as we struggle to develop our theoretical ideas ever more thoroughly and in close conjunction with an explicit articulation of learning goals for individual students.

187

REFERENCES

Karlin, S.: 1953. Some random walks arising in learning models, Part I. *Pacific Journal of Mathematics*, **3**, 725–756.

Larsen, I., Markosian, L. Z., and Suppes, P.: 1978. Performance models of undergraduate students on computer-assisted instruction in elementary logic. *Instructional Science*, **7**, 15–35.

Malone, T. W., Suppes, P., Macken, E., Zanotti, M., and Kanerva, L.: 1979. Projecting student trajectories in a computer-assisted instruction curriculum. *Journal of Educational Psychology*, **71**, 74–84.

Suppes, P.: 1964. Problems of optimization in learning a list of simple items. In M. W. Shelly II and G. L. Bryan (Eds.), *Human Judgments and Optimality*, New York: Wiley, 116–126.

Suppes, P.: 1967. Some theoretical models for mathematics learning. *Journal of Research and Development in Education*, **1**, 5–22.

Suppes, P.: 1972. Computer-assisted instruction. In W. Handler and J. Weizenbaum (Eds.), *Display Use for Man–Machine Dialog*. Munich: Hanser, 155–185.

Suppes, P., Fletcher, J. D., and Zanotti, M.: 1976. Models of individual trajectories in computer-assisted instruction for deaf students. *Journal of Educational Psychology*, **68**, 117–127.

Suppes, P., Groen, G., and Schlag-Rey, M.: 1966. A model for response latency in paired-associate learning. *Journal of Mathematical Psychology*, **3**:1, 99–128.

Suppes, P., Macken, E., and Zanotti, M.: 1978. The role of global psychological models in instructional technology. In R. Glaser (Ed.), *Advances in Instructional Psychology*, Vol. 1. Hillsdale, NJ: Erlbaum, 229–259.

Suppes, P., and Morningstar, M.: 1972. *Computer-Assisted Instruction at Stanford, 1966–68: Data, Models, and Evaluation of the Arithmetic Programs*. New York: Academic Press.

Teachers Handbook for Math Concepts and Skills. Computer Curriculum Corporation, 1993.

Author index

189

Subject index

Printed in the United States
By Bookmasters